# THE COSMIC COCKTAIL

## SCIENCE ESSENTIALS

Books in the *Science Essentials* series bring cutting-edge science to a general audience. The series provides the foundation for a better understanding of the scientific and technical advances changing our world.

In each volume, a prominent scientist—chosen by an advisory board of National Academy of Sciences members—conveys in clear prose the fundamental knowledge underlying a rapidly evolving field of scientific endeavor.

# THE COSMIC COCKTAIL

## THREE PARTS DARK MATTER

### KATHERINE FREESE

PRINCETON UNIVERSITY PRESS

*Princeton and Oxford*

Published by Princeton University Press, 41 William Street, Princeton, New Jersey 08540
In the United Kingdom: Princeton University Press, 6 Oxford Street, Woodstock, Oxfordshire OX20 1TW

press.princeton.edu

Jacket photographs: Metal cocktail shaker © Tony C. Real; stars in the constellation of Lyra © Oriontrail. Both images courtesy of Shutterstock. Cover design by Faceout Studio, Jeff Miller.

Library of Congress Cataloging-in-Publication Data
Freese, Katherine, author.
 The cosmic cocktail : three parts dark matter / Katherine Freese.
  pages    cm — (Science essentials)
 Includes bibliographical references and index.
 ISBN 978-0-691-15335-3 (hardback : alk. paper)
 ISBN 978-0-691-16918-7 (paperback)
 1. Dark matter (Astronomy)   2. Cosmology.   I. Title.
QB791.3.F75     2014
523.1′126—dc23                              2013042263

British Library Cataloging-in-Publication Data is available

This book has been composed in Garamond Premier Pro with DIN and Reforma Grotesk display by Princeton Editorial Associates Inc., Scottsdale, Arizona.

Printed on acid-free paper. ∞

Printed in the United States of America

*This book is dedicated to*
*three people who were a profound inspiration to me:*

To my parents Drs. Elisabeth and Ernst Freese,
who made seminal contributions to early research in genetics
and helped to create the field of molecular biology;
and to my PhD advisor Dr. David Schramm,
one of the founders of the field of particle astrophysics.

## The Cosmic Cocktail Recipe

3 oz. dark matter
7 oz. dark energy
1/2 oz. hydrogen and helium gas
3 thousandths oz. other chemical elements
5 hundredths oz. stars
5 hundredths oz. neutrinos
5 ten-thousandths oz. cosmic microwave background light
1 millionth oz. supermassive black holes

Shaken, not stirred.
Secret ingredient: dark matter

# CONTENTS

# PREFACE

"What is the Universe made of?" This question is one of the deepest unanswered mysteries in all of human existence. Solving the puzzle has been my life's work and is the hottest research topic in cosmology and particle physics today. The reason for the excitement is clear. The bulk of the mass in the Universe consists of a mysterious dark component, and its identity is on the verge of discovery. *The Cosmic Cocktail* tells the story of dark matter research and the race among groups of scientists to find the answer.

The book mixes my personal trajectory as a scientist with the rapid developments in the field. I begin by describing the path that led to my career as an astrophysicist. Then I turn to the phenomenal vision and achievements of physicists who preceded me and established the study of the Cosmos as a branch of science.

Modern cosmology began with Albert Einstein's breakthroughs in relativity theory in 1915. His work created a mathematical framework for physicists to study the structure and evolution of the Universe as a whole. Yet, even at the time of these tremendous developments in theoretical physics, the status of astronomical observations of the Universe remained quite primitive. As a result, a myriad of proposed models for the Universe stood on equal footing—a static Universe, an expanding Big Bang Universe, and a steady state Universe with continuous creation of matter from the vacuum of space. But only one of these could be correct.

In 1929 Edwin Hubble, using the telescope at Mt. Wilson Observatory in the mountains above Pasadena, California, studied the motions of galaxies and came to an astonishing conclusion. He noticed that, on average, galaxies are all moving apart from one another. To explain this observation, he postulated that they are being drawn apart as a consequence of the underlying expansion of the Universe. This notion of an expanding Universe initially disturbed Einstein, who, on aesthetic grounds, wanted the Universe to be static. However, science is ruled by data, and the astronomical observations were irrefutable. Eventually Einstein had to concede that Hubble was right. The Universe is expanding. The Big Bang model, in which the Universe started as a hot dense medium that has been subsequently spreading apart and cooling, became the standard model of cosmology.

Although we now know our Universe is of the Hot Big Bang variety, much of our Universe continues to be elusive. The nature of most of our Universe—the dark side—remains a mystery. In 1933, Fritz Zwicky at the California Institute of Technology in Pasadena first identified this puzzle. He noticed that galaxies in the Coma cluster were moving too rapidly to be understood by the gravitational pull of stars and other luminous matter alone. Zwicky postulated that most of the cluster must consist of some new, unknown component, which he named "dark matter." In the 1970s observations made by astronomers Vera Rubin and Kent Ford at the Carnegie Institution of Washington clinched the case for dark matter in galaxies. Identifying the nature of this missing mass has become the longest unsolved problem in modern physics. The turn of the millennium brought further revolutionary developments, including the discovery that there must be so-called "dark energy" driving the acceleration of the Universe. Scientists have since pieced together a cosmic inventory. The Cosmos can be imagined as a pie with three major slices: 5% ordinary matter, 26% dark matter, and 69% dark energy.[1] Nature can indeed be stranger than science fiction.

Astronomy has systematically whittled away at some of our deeply ingrained assumptions. The telescopes of Galileo and theories of Copernicus showed that Earth is not at the center of the Cosmos. In modern times, we have had to accept a further revision in the human worldview. The atomic matter we are composed of does not even come close to making up the bulk of the Universe. It is disquieting to learn that our type of matter is not even predominant—perhaps the ultimate Copernican principle. Yet whereas Galileo was sentenced to house arrest for his radical notions, modern civilization is instead supporting the search for scientific truth. Governments fund the research. Our society sees the technological advantages of science; but it also expresses a desire for a deeper understanding of our world.

Contemplating the vastness of the Universe makes humans feel small. Yet in some sense, our scientific knowledge has tamed the Cosmos. We know much of what it looks like all the way out to the *horizon*—the edge of our observable Universe. At a distance of 13.8 billion light-years from us, this is the farthest extent that light (or any other signal) could have traveled since the Big Bang. Farther out than that, nothing can surprise us: we are immune to anything that happens beyond the horizon, because its effects haven't had time to reach us in the age of the Universe. We understand most of the basic physics of everything out to this distance: the stars and the galaxies, as well as their evolution—everything except the nature of the dark side, that is, the dark matter and the dark energy.

In the century since Einstein first proposed his theories of relativity, the progress of our understanding of the Cosmos has been astounding. The twenty-first century has become the golden era of cosmology. Many age-old questions have been answered since the turn of the millennium: How old is the Universe? What is its shape? What is its total content? This book illustrates the measurements that answered these questions as well as the personalities behind the experiments. Yet answers to great questions bring new great puzzles. Now we have the dark side of the Universe to explain.

In this book I describe the unmistakable observational evidence for dark matter as the dominant mass in galaxies, in clusters, and in the Universe. Then I turn to the modern perspective, in which particle physics might provide the answer to the nature of dark matter. Some new particles, unlike any from our daily experience, might be tearing through the galaxy. With the possibility of billions of these exotic particles passing unnoticed through our bodies every second, we may hope to build sensitive detectors capable of resolving the identity of dark matter. Scientists have already found hints of detection in their experiments, and I tell their stories in these pages. The nature of dark matter is one of the greatest puzzles of modern science, and it is a puzzle we are on the verge of solving. But first, let me start with the story of how I became involved in dark matter and how this adventure has become the center of my professional universe.

# THE COSMIC COCKTAIL

# ONE

## The Golden Era of Particle Cosmology, or How I Joined the Chicago Mafia

I t was in a hospital bed in Tokyo that I realized I had to become a physicist. I was 22 years old. Although I had earned an undergraduate degree in physics from Princeton University 2 years before, I still felt unsure about what I wanted to do with my life. I had rushed to college at age 16, spent the first 2 years dating boys and the next 2 studying way too hard. After applying to graduate school, I was surprised when 12 of the 14 doctoral programs I'd contacted accepted me. But I needed a break. So I decided to travel the world.

Tokyo was the first stop. To earn money for the rest of the trip, I taught English for a while, then served drinks in a bar. It was as hostess in the bar that I learned to deflect men's advances and demand to be treated professionally— skills that later proved invaluable in the male-dominated physics world. Tokyo was my first experience of a big city, and I loved it. I stayed there for nearly 2 years.

As a side trip I took a boat to South Korea for sightseeing. There I found myself doubled over with stomach pain throughout most of the voyage. When I returned to Tokyo 2 weeks later, the pain suddenly became excruciating. I took myself to the first emergency room listed in my guidebook, at a hospital run by English nuns. I needed an emergency appendectomy. A surgeon operated immediately and discovered that my appendix had started to rupture. Afterward I lay in the hospital bed in agony, drifting in and out of consciousness. It was a semi-private hospital room, but for several days I didn't even realize I had a roommate. Once I started to feel better, I quickly became bored. The patient in the neighboring bed couldn't speak English, so I asked the nuns for a textbook to learn to speak Japanese. They looked at me as though they thought I was crazy. So instead I began reading the only physics book I had brought with me: *Spacetime Physics*[1] by Edwin Taylor and John Wheeler.

1

What an amazing subject and an amazing book. *Spacetime Physics* is all about Albert Einstein's theory of Special Relativity. It's built on two simple postulates: one, the speed of light is always constant, and two, the laws of physics don't depend on your state of motion—they're the same whether you're on a moving train or standing still on the platform. These two postulates immediately lead to bizarre phenomena. One is the twin paradox: if one of two identical twins speeds into outer space and then back to Earth, she will return younger than the twin she left behind. So in a way, relativity allows time travel to the future.

Reading *Spacetime Physics* was exhilarating. I spent the remainder of my week in the hospital reading the entire book and solving every problem at the end of each chapter. The book inspired me to want to learn more. I felt that if one elementary text could substantially alter my perceptions of the Universe, then I had to go back to school for a deeper understanding. If physics from the early 1900s could be so fascinating, what would the next millennium bring? After my release from the hospital, I hightailed it back to the United States, reactivated my acceptance from Columbia University, and headed to graduate school.

## New York City

I started graduate school at Columbia University intending to become an experimental high-energy physicist. During my first semester, I was a little side-tracked by many nights at Studio 54, the most famous nightclub in the world. It was exciting to dance alongside celebrities like Andy Warhol, Mariel Hemingway, Robert Duvall, and Diana Ross. I lived on 112th Street and ate daily at Tom's Diner next door, now well known from the *Seinfeld* television show. I even harbored thoughts of becoming an actor, like my two cousins. But the pull of physics was stronger. Despite all my partying, I passed my classes (barely). After a year and a half, I moved to Fermilab, the particle accelerator laboratory that at the time was the preeminent high-energy physics facility in the world.

## Fermilab: The Atom Smasher in the Prairie

Fermilab is built on a large farm in the middle of the prairie an hour west of Chicago. Buffalo graze on site, and visiting scientists are housed in transplanted farmhouses. The contrast with the urban environment of New York City was a jolt. The winter was extreme. The wind chill temperature dipped down to minus 80 degrees F. Any part of your body exposed for even 10 minutes would be frostbitten. Every night my car battery died, and the Fermilab crew had to come start it up again. When I tried to walk up the stairs into the High Rise, the central 15-story building where I worked, I couldn't make it against the wind. A few of us went to a theater in a nearby town to watch the movie *Reds,* which is set partly in Siberia. Halfway through the movie the theater owner had to refund our money, because he couldn't keep the theater warm enough. Outside the scenery looked exactly like the Russian steppes we had been watching in the movie. Siberia isn't impressive if you're in the middle of a good Fermilab winter. Then one day in April the temperature miraculously changed from minus 20 to plus 80. Spring lasted about 5 minutes during lunchtime.

I was one of only a handful of women at Fermilab. When I walked into the cafeteria, I felt the attention of the hundreds of male physicists. I was somewhat uncomfortable, but my heart soared at the prospect of working on such an impressive piece of machinery and discovering something new about the fundamental constituents of nature.

When I arrived at Fermilab, I joined a team working on an experiment to discover neutrino mass. Neutrinos are subatomic particles produced in radioactive decays and in nuclear reactions in the Sun. A member of our group, the head of the Fermilab theory division, promised to strangle one of the buffalo on site by hand if the experiment resulted in a discovery. Fortunately for the buffalo, we did not succeed. Instead the experiment placed limits on neutrino properties. The discovery of neutrino mass had to wait almost two more decades.

Fermilab's particle detectors were enormous; a student could gain familiarity with only a small part of the experiment. My first task was to check 1,000 phototubes to see whether they were working properly. This meant removing any input to them and then making sure they were recording zero signal, as they should. The test is known as measuring the dark current. I had to manually remove cables from the phototubes, check the signal, and then replace the cables. When I was done, my hands were bleeding from the effort. Today I joke that I switched from working on dark current to working on dark matter and dark energy.

Soon I realized that participating in a team of hundreds did not suit me. I enjoy having command of the entire project—from the initial idea, to the calculations, and finally to a publication—all on a quick turnaround time. I am a theorist by temperament, working with ideas, pen, paper, and computers. I prefer to work on my own schedule. When a senior experimentalist taking a night shift said to me, "You will learn to live on little sleep," I decided this field was not for me.

## Chicago: A New Zeitgeist in Cosmology

In the interest of getting into Chicago a few times a week, I wanted to take a class in the city. It was the beginning of October. I looked into acting classes at the theaters, but they had already started a month earlier. I am incredibly lucky that, of all the universities in America, the University of Chicago is one of the few where the fall semester begins in October. I signed up for a cosmology class twice a week in the early morning taught by Professor David Schramm.

Little did I know that I was putting my future in the hands of one of the giants in all of science, both literally and figuratively. He was a huge man, a wrestling champ as well as a leader of the field of modern cosmology. He was a pioneer in the emerging field known as particle astrophysics. In 1968 he had been a finalist in the Olympic trials for Greco-Roman wrestling. We nicknamed him Schrambo. Dave pushed the limits both in his work as an astrophysicist and in his daily life. He was an expert skier, though not particularly graceful. With his enormous leg strength, he powered his way down double black diamond runs.[2] He seemed to simply shove the moguls out of the way.[3]

Twice a week I drove the hour from Fermilab to Schramm's early morning class at the university. I loved the course and found myself skipping my work at the experiment to stay behind (in the farmhouse where I lived) to read cosmology. Here the book with the greatest impact on me was Steven Weinberg's *Gravitation and Cosmology*.[4] Though I was merely auditing, I went ahead and took the exams in the class. The day before the midterm, I saw the exam sitting on the secretary's desk. Despite the temptation, I didn't look at the problems. I aced the test nonetheless, and as a consequence, Dave asked me if I would like to work on a project with him. Had I cheated, I would never have known whether his confidence in me was misplaced. I considered his proposal of a collaboration and then asked him if I could switch to become his graduate student. He immediately agreed, and that is how my career in cosmology began (Figure 1.1). I was lucky to get into the new field of particle astrophysics in its infancy, at a time when even simple ideas could have a big impact.

One of the benefits of becoming a theoretical physicist was the opportunity to visit the Aspen Center for Physics in Colorado. George Stranahan, physicist and heir to the Champion Spark Plug fortune, founded the center in 1962. There physicists spend long blocks of uninterrupted time thinking and brainstorming together. Important ideas have emerged from discussions among experts in different specialties who would never have met outside the center. I've participated in many of the summer and winter workshops. I enjoy the sports activities, including biking up the Rockies and skiing down them.

FIGURE 1.1 A picture with my PhD advisor David Schramm behind the Astronomy and Astrophysics Center at the University of Chicago in 1983.

Ever since a bee stung my leg while I was traveling downhill on my bicycle at 50 miles an hour, I ride the bike up the mountain and take the bus back down. Dave tried to get me to go mountain climbing with him, but I declined because of stories I'd heard from his previous student. They were roped together, when the student saw Dave lose control above him and come hurtling down. They were both barely hanging on when a helicopter rescued them. As they were being airlifted, Dave started haggling with the pilot about the bill.

Unfortunately, Dave Schramm's risk-taking style eventually cost him his life. He flew his own plane, which he named Big Bang Aviation. On the way to Aspen the engine failed, and Dave was forced to land next to the highway. Apparently the wing snagged a tree, the plane flipped over, and sadly, Dave died. Another close friend of mine (and former fiancé), Josh Frieman, had originally intended to be a passenger on the flight. Fortunately, he overslept, or I would have lost another important person in my life. My last image of Dave Schramm is on the final ski run of the day down Aspen Mountain. He had found a patch of remaining powder under a ski lift already shut down for the night. He yelped with excitement as he carved his way down the mountain, and that is the same approach he took to doing science.

Chicago felt like the center of the world for particle astrophysics. Dave Schramm led the Center for Astronomy and Astrophysics at the University and also created the Center for Particle Astrophysics at Fermilab. He hired

Michael Turner, who quickly rose to become another of the leaders in cosmology. Michael, then an assistant professor, was the other important person in my early career. He taught his students to do back-of-the-envelope estimates that are so crucial to determining whether an idea is worth pursuing. After we had done the calculations, he showed us the mechanics of putting a paper together quickly. Michael had long hair, wore humorous T-shirts, and often treated students to lunch. When I was in between offices, Michael let me use his.

The University of Chicago became a cutting-edge center for students in particle astrophysics. Those of us who were trained there came to be known as the Chicago Mafia.

## Dunkle Materie: The Dark Enigma

After obtaining my PhD, I went to Harvard as a postdoctoral fellow, and it was there that I started working on unraveling the dark matter problem. This mystery began in the 1930s and remains one of the grandest unsolved problems in all of science. The rapid motions of stars and gas in galaxies and clusters of galaxies imply the existence of a new massive component exerting a powerful gravitational pull. This dark matter is now known to dominate the mass of galaxies and clusters, as we'll see in Chapter 2. We call it "dark" because it does not give off light and cannot be seen in telescopes. Its nature is mysterious, but its existence is certain. Using the recent technology of gravitational lensing, scientists have mapped out the dark matter inside typical galaxies and shown that it extends out to hundreds of thousands of light-years.

Most people believe the predominant constituent of our surroundings to be the ordinary atomic matter of our daily experience: the chairs we sit on, the walls of our rooms, the air that we breathe, the planets and stars. Yet over the past few decades it has become clear that all these objects add up to only 5% of the content of the Universe. All atomic matter, which is made of neutrons and protons (or at a more fundamental level, quarks) and electrons, only constitutes a small portion of the matter and energy in the Universe.

A greater part of creation is dark matter, which makes up most of the mass of the Universe. Scientists believe the dark matter is made of a new type of fundamental particle—not neutrons, protons, or quarks, but something altogether different. Billions of these particles may be passing through our bodies every second, yet typically they go right through us without hitting any of our nuclei. Because they interact weakly with matter, they are very difficult to identify in detectors. And so the nature of this material has remained a deep mystery.

In the 1980s, the scientific community was divided into two camps debating the identity of dark matter. Many of the more traditional astronomers believed it consisted of faint stars or substellar objects. Although bright stars couldn't solve the problem, it was certainly possible that huge numbers of faint stars could exist, just beyond the limits of detection in telescopes. However, even at that time there was a new Zeitgeist in cosmology. Scientists originally working on particle physics were beginning to have a major impact on astronomy. This second camp of astroparticle physicists was proposing the idea of new fundamental particles as the likely origin for most of the mass in the Universe. This disagreement became the battle between the MACHOs and the WIMPs. Here MACHO stands for Massive Compact Halo Object (that is, some type of stellar object), and WIMP stands for Weakly Interacting Massive Particle, the most likely particle candidate for dark matter. I played a role in both sides of this debate.

### WIMPs at Harvard

In the mid-1980s I was fortunate to be in the right place and the right time to make early contributions that started the hunt for dark matter particles. My collaborators and I sat around a table at the Harvard / Smithsonian Center for Astrophysics and made proposals for ways to search for WIMPs. In the quarter of a century since we wrote down our ideas, the experimental race for dark matter detection has been on.

At the Large Hadron Collider at the European Organization for Nuclear Research (known by its French acronym, CERN)[5] in Geneva, for instance, physicists accelerate two proton beams in opposing directions around a 27-kilometer (17-mile) long ring and smash them together at tremendously high energies. A major goal of this enterprise is the hunt for dark matter. Laboratory "direct detection" experiments are designed to detect traces of dark matter particles from the Galaxy striking nuclei in the detectors. These experiments are recording data deep underground, in abandoned mines and Alpine tunnels. "Indirect detection" experiments aboard balloons, at the South Pole, or mounted on satellites in space are searching for signs of dark matter annihilation products. Now even dark matter detectors made of DNA have been proposed.

Some of these experiments are reporting anomalous signals and even claims of discovery. Theorists are feverishly trying to explain all the results. The current situation is both exciting and perplexing.

The field of modern dark matter cosmology was created by a surprisingly small number of individuals. The experimental groups are modest in size, typically consisting of a dozen people. Perhaps it is for this reason that a disproportionate number of women have gone into dark matter studies. An individual can make a mark without jostling with the crowd and without fighting preconceptions about what the leader of a large group should look like.

The personalities of the leaders of these experiments are strong and can create intense conflicts. Yet at the same time the dark matter community is full of vibrant friendships. I am close to many of these people, and in the following pages I tell their stories as they intersect with mine. The tendency is to depersonalize scientific discoveries and to attribute them in the end to one person. But in reality, it is the collaboration of many people with sympathetic scientific views and outlooks that solves problems—and the hunt for dark matter is no exception. Through this community effort, I believe that the unraveling of the dark matter problem is at hand.

# TWO

# How Do Cosmologists Know Dark Matter Exists?
## *The Beginning of the Dark Matter Story*

Fritz Zwicky was an irascible fellow—but a brilliant scientist. Raised in Switzerland, he moved to the United States in 1925. He spent most of his professional life at the California Institute of Technology in Pasadena. His creativity and contributions to astronomy were tremendous. In the 1930s, Zwicky studied the Coma Cluster, a rich cluster containing thousands of galaxies. He noticed that the individual galaxies were moving surprisingly fast—far more rapidly than could be explained by the gravitational pull of the stars observed in the cluster. In fact, with their colossal speeds, they should have escaped from the cluster entirely.

To explain this discrepancy, Zwicky postulated the existence of a new massive component of the Coma Cluster. He speculated that some unidentified matter must be pulling on the galaxies and speeding them up. Since this material does not produce any light and is invisible to telescopes, he called it "dunkle Materie," the German for "dark matter." The word "dark" simply refers to the fact that this missing mass does not produce light of any kind; it does not shine. Zwicky had no idea what constituted the dark matter, but he argued for its existence based on its gravitational effects on visible matter. He estimated that there must be 200 times as much dark as luminous mass in the Coma Cluster to explain his observations. Since that time, the ratio has been revised to closer to 10. But the basic notion that some sort of mysterious dark component dominates the mass of galaxies and clusters persists. So began the puzzle of dark matter in the Universe, the longest outstanding unsolved problem in all of modern physics.

At Caltech Zwicky became a legend (Figure 2.1). He dubbed his colleagues "spherical bastards." "They are spherical," he said, "because they are bastards every way I look at them." He must have been quite intimidating to his students! An avid alpine climber and skier, he built himself a ski jump near the

FIGURE 2.1 Fritz Zwicky (1898–1974). *Courtesy of the Archives, California Institute of Technology.*

telescope on Mount Wilson above Pasadena and used it in the winter. Some of his ideas were quite extravagant. "Nudge Mars closer to the Sun and see if it becomes habitable," he suggested. "Nudge the Sun itself. Send it and all its gravitationally bound bodies, including Earth, toward a star with habitable planets so we might one day colonize other solar systems." Who knows? Many of Zwicky's other ideas were incredibly prescient and in time proved to be correct. Maybe a traveling Solar System is in our future.

Zwicky made other major contributions to astrophysics. He proposed the existence of neutron stars,[1] which are the final states of the stellar evolution of stars more massive than three Suns. Once these have exhausted all their nuclear fuel, they collapse to become small dense objects with their entire mass condensed inside a few kilometers.[2] By now thousands of neutron stars have been discovered. Zwicky also postulated that, en route to becoming neutron stars, the precursor stars blow up as supernovae, among the brightest, most energetic events in the Universe. He discovered the first supernova and went on to find another 120 of these objects. We'll see in Chapter 9 that studies of supernovae have led to another major discovery in astronomy—the existence of dark energy.

The dark matter problem identified by Zwicky still remains one of the deepest unsolved problems in modern science. Since his time, the evidence for dark matter has become indisputable. Though dark matter does not shine, a host of different types of observations identify its gravitational effects on visible matter throughout the Universe. Its nature remains mysterious, but dark matter clearly constitutes the dominant component of galaxies and clusters. The remainder of the chapter begins by describing the basic shape of galaxies, including their massive dark matter structures, and then turns to the observational evidence for the existence of dark matter in the Universe.

### What Do Galaxies Look Like?

In the 80 years since Zwicky first proposed the existence of dark matter, astronomers have converged on a basic understanding of the shapes and dynamics of galaxies. Let's start with a description of our own Milky Way, the galaxy we inhabit.

When we look at the night sky, we see a band of stars that we think of as the Milky Way. In reality, what we are seeing is only a small fraction of the total Galaxy. The Milky Way as a whole weighs roughly a trillion solar masses—a trillion times as much as the Sun. Yet the stellar material, the objects that astronomers can observe with telescopes, constitutes only 5% of this total mass. Almost all of the stars are on a flat central plane known as the disk. This disk, shaped much like a Frisbee, rotates around the center of the Galaxy at roughly 250 kilometers per second, or 560,000 miles per hour. Figure 2.2 shows the flattened disk structure that contains the Galaxy's stars, including our Sun.

A supermassive black hole resides at the center of the Galaxy, weighing about 4 million solar masses.[3] The black hole occupies a small central region about as large as the diameter of Neptune's orbit. Though the central black hole has a strong gravitational pull on nearby objects, here on Earth we don't need to worry about being swallowed up.[4] Our Sun is too far away from the black hole for us to notice its gravity at all. The Galaxy as a whole weighs a million times as much as its central black hole.

Spreading out from the center of the Galaxy along the disk are long spiral arms of stars, much like a pinwheel. Our Sun resides along one of these spiral arms, about 25,000 light-years away from the center. A light-year is the distance light can travel in a year, about 10 trillion kilometers (6 trillion miles).

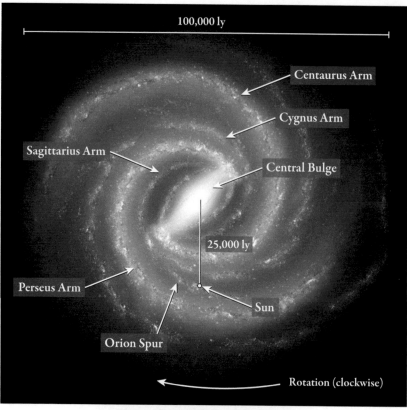

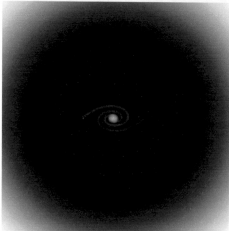

FIGURE 2.2 The Milky Way. (Top) The pinwheel shape of the spiral arms of the disk of the Galaxy is shown in the drawing. The central bright region is the Galactic Bulge, a central region of tightly packed stars. At the very center of the Galaxy is Sagittarius A*, a strong radio source affiliated with a black hole weighing 4 million times as much as our Sun. The Sun is located about 25,000 light-years away from the Galactic Center. (Bottom) Artist's rendition of a galaxy, showing the spiral structure of the disk as well as the much larger spherical dark matter halo. *(Top) NASA/JPL-CalTech.*

Surrounding the flat stellar disk is a huge spherical massive globe, known as the Galactic halo, which contains most of the mass in the Galaxy. The halo is essentially dark; it contains very few stars. Instead, because of its gravitational influence, we know it consists mostly of dark matter. One can think of the stellar disk as a Frisbee, contained in a much larger basketball filled with dark matter. The word "halo" can be misleading. There is absolutely no connection to the ring-shaped halos hovering over the heads of saints in medieval paintings. Galactic halos are not donut shaped; they are spheres containing dark matter.

Galaxies with disks and spiral structure, such as the Milky Way, are known as spiral galaxies. In contrast, elliptical galaxies, constituting roughly 10% of observed galaxies, are football-shaped objects with old stars throughout. All galaxies, including both the spirals and ellipticals, share two features: they have supermassive black holes at their centers, and they have giant dark matter halos.

### Observational Evidence for Dark Matter

After Fritz Zwicky first introduced the notion of dark matter in clusters in the early 1930s, others wondered about the individual galaxies themselves. Might these smaller structures have a large dark matter component as well? Here we examine the mounting evidence for dark matter in galaxies over the past 80 years.

### Rotation Curves of Galaxies

So-called "rotation curves" can be used to weigh galaxies and other astronomical systems. Such rotation curves provided the first evidence for large spherical dark matter halos in galaxies. As a simple analogy, let's start with the rotation curve of the Solar System. Figure 2.3 shows a plot of this rotation curve—the relationship between the average speeds of the planets and their distances from the Sun. In the figure, the Sun is located on the left axis (at $r = 0$). Moving to the right in the plot corresponds to moving farther away from the Sun. The closer a planet is to the Sun, the faster it moves. Mercury, which is closest to the Sun, completes a full orbit in only 88 days. The second-closest planet, Venus, takes 225 days. As we all know, the time for Earth to circle the Sun is one year. For comparison, Jupiter takes almost 12 Earth-years to complete one orbit. The planets farthest out from the Sun move the most slowly and take the longest to make it all the way around.

Eight planets are plotted in the figure. Pluto, though originally classified as a planet, is not shown. It has been demoted to a Kuiper belt object. Pluto,

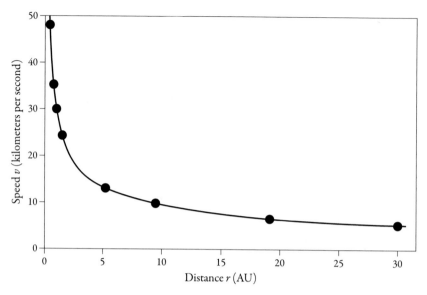

FIGURE 2.3 The rotation curve of the Solar System: average speeds of the planets versus distance from the Sun. The Sun (not shown in the plot) is located at $r = 0$; the planets are at distances as indicated in astronomical units (AU). One AU is the distance between Earth and the Sun, roughly 100 million kilometers. The farther out a planet is from the Sun, the more slowly it revolves in its orbit.

which weighs roughly a fifth as much as the Moon, is composed primarily of rock and ice. It follows a chaotic elliptical orbit quite far out from the Sun.

Based on the rotation curve of the Solar System, we can determine the mass that is causing these planetary motions. In this case, it is the mass of the Sun, and its gravitational pull keeps all the planets in orbit. The laws of physics discovered by Isaac Newton explain the shape of the rotation curve. Newton's laws tell us that the orbital speed scales as $v = \sqrt{GM/r}$. Here $r$ is the distance from the Sun, $M$ is the mass of the Sun, and $G$ is Newton's constant. More accurately, the mass $M$ should be the total mass of the system interior to the orbit of any object. Because the Sun is by far the most massive object in the Solar System, we can take $M$ to be the mass of the Sun. The speeds of the planets in Figure 2.3 follow the drop-off with distance exactly as predicted by Newton's laws. This relation implies that at four times the distance, the speed is half as large (lower by $\sqrt{1/4}$). In other words, an object that is on an orbit (around the Sun) four times farther out than Earth would be moving half as fast. The equation shows that speeds of the planets depend not only on their distances from the Sun but also on the Sun's mass. If the Sun were four times as massive, the speeds of all the planets would double (growing by $\sqrt{4}$).

FIGURE 2.4 Tycho Brahe (1546–1601). He lost his nose in a
duel and wore a gold-and-silver replacement. Brahe is known
for his studies of planetary orbits.

By measuring the speeds of objects orbiting around the center, we can
determine the amount of mass interior to their orbits. For example, if measurements of Jupiter's speed had been found to be significantly faster, we would
have to reevaluate the mass inside Jupiter's orbit. One possibility would be
that we might have misjudged the Sun's mass. Or, because the totality of all
the mass interior to a planet's orbit determines its speed, another interpretation would be possible. We might have missed some mass somewhere in the
Solar System between the Sun and the orbit of Jupiter. We can use the speeds
of orbiting objects to obtain estimates of the mass supplying the gravitational
pull on the objects.

The Danish nobleman Tycho Brahe, who lived from 1546 to 1601, made
some of the most important and accurate measurements of these planetary
orbits (Figure 2.4). In addition to his pioneering work in astronomy, he is

known for his escapades. He lost his nose in a duel and thereafter wore a gold-and-silver replacement. As the story goes, he died of a bladder infection after a dinner with the king. Although he drank a great deal at the banquet, he was too polite to stand up from the table to urinate while the king was still seated. Brahe died 11 days later. Examination of hair from his beard in 1901, however, gave rise to alternate theories of his demise. The hair contained toxic levels of mercury. Some thought he might have been poisoned. Soon after his death, his notebooks were missing. There is speculation that his fellow astronomer Johannes Kepler might have stolen Brahe's work. Perhaps Kepler's laws of planetary motion would otherwise be known as Brahe's laws. After 1901, Brahe's remains were placed inside a tin box the size of a child's coffin in a tomb in Prague's Old Town Square. Then in 2010, his body was exhumed from his grave in Prague. A team of scientists measured the amount of mercury in it using a CT scan, an x-ray analysis, and a neutron activation analysis. In 2012 the team concluded that the levels of mercury found in his body would not have been enough to kill him. Most likely Brahe died of a burst bladder after all.

Rotation curves can be used to weigh galaxies. Similar to the rotation curve of the Solar System, astronomers obtain rotation curves of galaxies. They measure the speeds of stars and gas at various distances from the center of a galaxy. Figure 2.5 indicates the path of a star circling a galaxy. The star's speed depends on the gravitational mass interior to its orbit. Larger mass causes faster orbits. Astronomers plot the relationship between the speeds of the stars and their distances from the center to determine the amount of mass in the galaxy.

Based solely on the gravity of the luminous matter in galaxies—the stars and other light-emitting material—the rotation curves should fall off with distance from the center, just as in the case of the Solar System. The lower curve in Figure 2.6 shows the expected shape of the curve if the stellar material were the only mass in the galaxy. The initial rise of the curve at the far left of the plot, just outside the Galactic Center, is easy to understand. Unlike the case of the Solar System, not all the luminous mass of galaxies is at their centers. Thus the orbital speeds, even in the absence of dark matter, should have this initial rise with distance from the center, as material outside the center comes into play. However, once all the luminous mass is encompassed, then the speed should fall off again. Based only on the stellar material in galactic disks, the gas and stars far out from the center should be moving more slowly than those that are closer in.

However, astronomers found something radically different. In 1939, for his doctoral dissertation, Horace Babcock studied motions of stars in the nearby Andromeda galaxy. Though he had a minimal amount of data, he plotted a rotation curve of these stars. Based on this information, he estimated

FIGURE 2.5 The path of a star in a circular orbit around the center of a galaxy. The orbital speed $v$ is determined by the amount of mass interior to the orbit at radius $r$. *NASA/JPL-Caltech/STScI.*

the amount of mass inside the galaxy required to speed up the stars to their observed velocities. The rotation curve looked inconsistent with the one predicted from the stellar material alone. He found evidence of a "flat" rotation curve.[5] The speeds of objects were the same, no matter how far out they were from the center. We'll discuss the interpretation of this behavior in a minute. His work was not persuasive, as he had only four data points with large errors. Even with these minimal data, he was able to use this technique to deduce the mass of the nearby Andromeda Galaxy to close to the correct value. Andromeda weighs about a billion times as much as the Sun—about the same as the Milky Way. Yet Babcock attributed the flatness of the rotation curves as most likely due to absorption of stellar light or other effects rather than to dark matter. Over the next 40 years, researchers attempted to find rotation curves of other galaxies with varying degrees of success.

In the 1970s, Vera Rubin and Kent Ford of the Carnegie Institution of Washington (in Washington, D.C.) made an extremely important contribution.[6] They succeeded in obtaining accurate measurements of rotation curves of many galaxies. It was their work that led to consensus in the physics community that galactic rotation curves are flat. The upper curve in Figure 2.6, with bars indicating the data, shows a flat rotation curve (the example shown dates

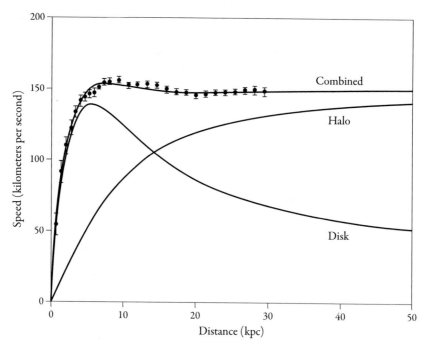

**FIGURE 2.6** The rotation curve of galaxy NGC 3198. Radial distances are measured in units of kiloparsecs (kpc), or thousands of parsecs; a parsec corresponds to roughly 30 trillion kilometers (19 trillion miles). The dots with bars show the observational data, whereas the solid curves show the calculated contributions from the stellar disk, the dark matter halo, and the two combined. Dark matter is required to explain the data. *Redrawn from van Albada, T. S., J. N. Bahcall, K. Begeman, and R. Sancisi. 1985. "Distribution of Dark Matter in the Spiral Galaxy NGC 3198." Astrophysical Journal 295: 305.*

from 1985). Objects orbiting the galaxy continue to move rapidly even at great distances from the Galactic Center. As mentioned above, based on the known stellar matter in the galactic disk, the speeds should have fallen off rapidly. Yet they did not. The most plausible interpretation is that the galaxy must contain additional mass. The difference between the observed rotation curve and the one predicted by disk stellar material is then due to dark matter. The intermediate curve in Figure 2.6, labeled "halo," indicates the additional material that must exist to explain the data. Adding a spherical halo component made of dark matter resolves the discrepancy between the two rotation curves. It was the work of Rubin and Ford that clinched the case for dark matter in galaxies. Their observations persuaded astronomers that dark matter must exist. Rubin has been awarded a National Medal of Science for her work, and the two deserve a Nobel Prize for this discovery (Figure 2.7).

**FIGURE 2.7** Vera Rubin, Carnegie Institution of Washington.
*Astronomical Society of the Pacific / Mark Godfrey.*

Later work of many others continued to confirm these results. Morton Roberts and Robert Whitehurst obtained galactic rotation curves by measuring the motion of hydrogen gas. Their results were crucial, because this gas can be observed out to much greater distances than can stellar light. Additional important contributions include those of Ken Freeman.[7] Rotation curves have been measured for hundreds of galaxies, and they are all observed to be flat.[8]

Even the orbit of our Sun around the Galactic Center is affected by dark matter. Though we are quite close to the center (relative to the distances measured by Rubin and Ford and others), still we are sped up by the dark matter interior to the Sun's orbit. The Sun is moving around the Galactic Center at roughly 250 kilometers per second, whereas the speed should be only 160 kilometers per second in the absence of a Galactic halo. Again, the discrepancy between the expected and observed speed of the Sun is due to dark matter. The Sun is sped up by the dark matter interior to its orbit around the Milky Way.

All measured rotation curves of galaxies are flat. The most likely explanation is that galaxies consist predominantly of dark matter. Alternative ideas have been proposed to explain the rotation curves, such as Modified Newtonian Dynamics

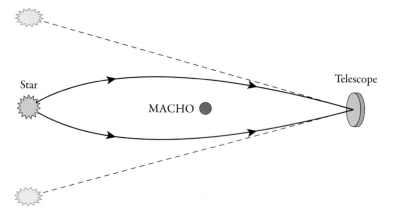

**FIGURE 2.8** Gravitational lensing: a Massive Compact Halo Object (MACHO) bends the light of a distant star seen by a telescope on Earth. Whereas the actual star is located directly behind the MACHO, the telescopes see what appear to be two distorted images at different locations than the real star. Any intervening mass between the star and the telescope would have the same lensing effect.

(MOND). Later in the chapter, we'll see, however, that these alternate theories have serious flaws. The consensus in the cosmology community is that the mass of all galaxies, including our own Milky Way, must be made of dark matter.

### Einstein's Gravitational Lensing

In the past decade, gravitational lensing—the bending of light by a massive object—has become a powerful tool for locating dark matter in the cosmos. This phenomenon is a consequence of Albert Einstein's theory of General Relativity.

Figure 2.8 illustrates the basic setup for gravitational lensing. A telescope on Earth observes a star (to the left in the picture). In between the star and the telescope, the light encounters some intervening material or massive body. In the picture, this mass is labeled as a MACHO (Massive Compact Halo Object)—but it could be any mass. The light-emitting star is known as the source, whereas the intervening matter is the lens. According to Einstein's relativity, the lensing mass distorts the path of the light en route to Earth by pulling it toward the mass. The light from the source star is focused, or lensed, onto the telescope by the dark massive object. As a result, if we assume the light has traveled in a straight line, we will misidentify the star's location. Though its real position is directly behind the MACHO, its apparent position will be off

FIGURE 2.9 A picture of lensed University of Michigan students (through the bottom of a drinking glass). (Top left) Unlensed; (top right) weakly lensed; (bottom left) strongly lensed; (bottom right) very strongly lensed, producing many images. *Alejandro Lopez.*

to the side. In fact, in the example in the figure, two identical images of the star will appear on either side of the MACHO image.

Soon after Einstein predicted this phenomenon in 1915, astronomers looked for the bending of light by the Sun. They waited for a solar eclipse, so that the Sun's emission wouldn't dwarf the stars they were studying. They made precise measurements of stars behind the Sun and confirmed that the light path was indeed curved by the mass of the Sun, exactly as predicted by General Relativity.

An analogy illustrates a variety of possible lensing phenomena. Imagine looking at the world through the bottom of a drinking glass. Everything becomes distorted. Figure 2.9 shows some funny images of University of Michigan summer students seen through this lens. Depending on the shape of the glass, you may see several images of the same student. The photos show that

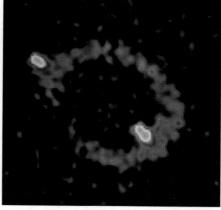

FIGURE 2.10 Jacqueline Hewitt, Director, Kavli Institute for Astrophysics and Space Research, Massachusetts Institute of Technology, together with the Einstein Ring she discovered in 1987 (object MG1131+0456). *(Left) Paul His, MIT; (right) NRAO/AUI.*

images can be magnified, distorted, or multiplied. Gravitational lensing can produce the same phenomena.

In some instances, three, four, or even more images of the same source may appear because of gravitational lensing. In the rare case where the lensing mass is lined up perfectly between the source and the telescope, an entire ring of images could appear, known as an Einstein Ring. With this setup, the light can travel around the intervening mass in any direction. Imagine drawing a circle around the lens; the light could pass through any point on that circle. Jacqueline (Jackie) Hewitt at the Massachusetts Institute of Technology in Cambridge was the first to discover an Einstein Ring in 1987. She studied radio emission of a bright distant object known as a quasar and found the image shown in Figure 2.10. A galaxy between the quasar and Earth was responsible for producing this ring. For this discovery, she was named "one of the top scientists under 40" by *Time Magazine*.

More generally, the number of images and their degree of distortion depend on the mass distribution of the lens as well as the alignment of the source, lens, and telescope. Detailed information about the lens can be extracted from the observations with the help of a little mathematics. In some cases, the resolution of the telescope is not good enough to distinguish the multiple images from one another. Yet the gravitational lensing still causes the source star to look brighter because of the additional light focused onto the telescope. As a clump of dark matter passes in front of the background source star, it can cause the light from that star to brighten for a period of a few days to a few months until the clump has moved on. This brightening of the image is known as micro-lensing. All the variants of gravitational lensing can be used to identify the mass of the intervening material.

Gravitational lensing has become an immensely powerful tool for locating mass in the Universe. It has provided astronomers with a great deal of knowledge about the matter in galaxies and clusters—most of it dark matter. Light feels the gravitational pull of any matter, whether it is bright or dark. The images of distant sources are bent, distorted, magnified, or brightened, depending on the configuration of the intervening mass. The background source can be thought of as a flashlight that shines on intervening dark matter and allows astronomers to see it, if only by its gravitational action on the passing light.

Using gravitational lensing, astronomers have obtained beautiful observations of the dark matter in clusters of galaxies. NASA's Hubble Space Telescope (HST) has produced spectacular images of the distant Universe. Launched into orbit by a space shuttle in 1990, HST has been one of NASA's great success stories. Though its mirror was originally found to be faulty, it was repaired by a servicing mission in 1993. A group of scientists from Bell Labs (Murray Hill, New Jersey) led by Tony Tyson used HST to study a large remote cluster (CL0024+1654) containing hundreds of galaxies. The cluster lensed the light from a more distant bright galaxy. This background galaxy, rather than a background star, served as the flashlight illuminating the cluster. The cluster produced multiple images of the source galaxy. Based on this lensing, Tyson and collaborators were able to reconstruct the underlying dark matter of the cluster.

Figure 2.11 shows a computer-generated image of the dark matter in the cluster. The sharp peaks are the individual galaxies. A large central hump of dark matter resides between the galaxies. The central region of the cluster contains a huge amount of mass, 250 times as much as the individual galaxies. Most of this dark matter is distributed smoothly, in a region of the sky about 2 million light-years across. We can think of the cluster as a mountain range of dark matter, with the galaxies protruding as individual peaks. This remarkable picture underscores the fact that, as we look out into the night sky and see the stars in galaxies, we are really only looking at the tip of the iceberg. The bulk of the mass consists of the mountain of dark matter that we are incapable of seeing directly by eye.

Gravitational lensing can also produce another effect: the distortion of bright background objects into long thin arcs. Figure 2.12 shows an image of

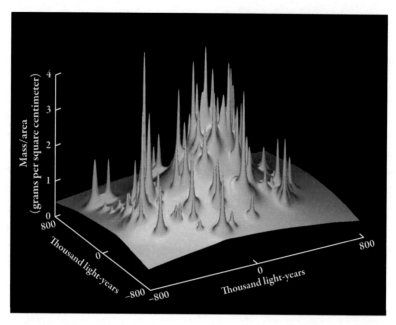

**FIGURE 2.11** (A color version of this figure is included in the insert following page 82.) Computer reconstructed image of the mass distribution in galaxy cluster CL0024+1654, based on data from the Hubble Space Telescope. This massive cluster gravitationally lensed the light of a more distant bright galaxy, producing multiple images of the source galaxy and allowing scientists to reconstruct the hidden mass inside the cluster. The peaks in the image are galaxies; the bulk of the mass consists of the central mountain made of dark matter in between the galaxies. *From Tyson, J. A., G. P. Kochanski, and I. P. Dell'Antonio. 1998. Astrophysical Journal Letters 498: L107.*

the rich cluster Abell 2218 taken by the Hubble Space Telescope. The background galaxies, which are 10 times farther away than the lensing cluster, have been sheared into thin arcs because of the lensing action of the intervening cluster. This image of Abell 2218 is remarkable in that it has seven cases of multiple imaging, in which the background galaxy is seen multiple times. The abundance of lensing images in this cluster has been used to make a map of its mass distribution. The dark matter content is enormous.

These images of gravitationally lensed clusters were made by the Hubble Space Telescope. Its successor, the James Webb Space Telescope (JWST) is an $8 billion project, due to be launched in 2018.[9] JWST's cameras will be able to obtain images of the faintest and most distant objects ever observed. With its sensitivity to infrared light and fantastic resolution, JWST will look even far-

FIGURE 2.12 Galaxy cluster Abell 2218 acts as a powerful lens. It is so massive that its gravity bends, magnifies, and distorts the light from galaxies lying behind the cluster into elongated arcs and multiple images. This image is from data taken with the Hubble Space Telescope. *NASA, Andrew Fruchter and the ERO Team [Sylvia Baggett (STScI), Richard Hook (ST-ECF), Zoltan Levay (STScI)] (STScI).*

ther out in space and backward in time than HST can.[10] It should give us views of the Universe back to 100 million years after the Big Bang.

### Hot Gas in Clusters

Another piece of evidence for dark matter comes from the existence of hot gas in clusters. Figure 2.13 shows images of the Coma Cluster, the same cluster studied by Zwicky. It is one of the richest clusters known, containing more than 1,000 member galaxies. Each of these galaxies houses billions of stars. The cluster is so big that it takes light millions of years to get from one side of it to the other.

Images of Coma have been taken by a variety of telescopes observing the light at different wavelengths. The top panel in Figure 2.13 shows a picture of the visible light coming from Coma. We can see the many galaxies it contains. In contrast, the bottom panel shows a picture of x-rays coming from the same Coma Cluster. The shades in the picture represent different intensities of x-ray emission. The two panels are not on the same scale: the x-ray picture shows only the central-most parts of the cluster. The x-ray images were taken with the European Space Agency's ROSAT satellite. At the lower right of the x-ray image, we can see a fainter group of galaxies that are in the process of merging with Coma's bright central cluster. The bottom panel shows

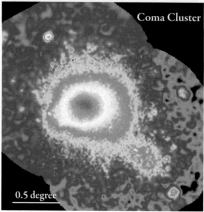

**Coma Cluster**

0.5 degree

FIGURE 2.13 (A color version of this figure is included in the insert following page 82.) Galaxy cluster Coma provides evidence for dark matter. The x-rays in the image on the bottom are produced by hot gas, which would have evaporated from the cluster without the gravity provided by an enormous dark matter component in the cluster. (Top) Optical image. (Bottom) X-ray image. The two images are not on the same scale; the x-ray image focuses on the central region of the cluster. *(Top) NASA, ESA, and the Hubble Heritage Team (STScI/AURA); (bottom) ROSAT/MPE/S. L. Snowden.*

that the cluster is pervaded by hot x-ray-emitting gas, which has a temperature of 10–100 million Kelvins (K). There must be a huge additional massive component that keeps this gas gravitationally bound to the cluster. Without dark matter, the hot gas would evaporate and escape from the central region. There would not be enough gravity to hold it in. It seems difficult to evade the

**FIGURE 2.14** (A color version of this figure is included in the insert following page 82.) The Bullet Cluster, a merger of two clusters each containing dozens of galaxies, gives striking confirmation of the existence of dark matter. In the color version of the figure, the dark matter from lensing measurements is shown in blue; the x-ray gas composed of atomic matter is shown in red. The separation of the two components occurs because the atomic gas decelerates when it collides at the center, but the collisionless dark matter passes right on through. The existence of these two independent components is exactly as predicted in dark matter theories. The name "Bullet Cluster" refers to the striking illusion that one of the clusters looks like a bullet piercing the other. *(X-ray) NASA/CXC/CfA/M. Markevitch et al.; (lensing map) NASA/STScI, ESO WFI, Magellan/U. Arizona/D. Clowe et al.; (optical) NASA/STScI, Magellan/U. Arizona/D. Clowe et al.*

conclusion that even the most enormous rich clusters must have a substantial dark matter component.

## The Bullet Cluster: Death to Alternative Theories of Gravity?

The year 2006 saw the discovery of the remarkable Bullet Cluster. This object actually consists of two separate colliding clusters, each containing dozens of galaxies. These clusters are in the initial stages of merging, a process that could take 100 million years. Scientists observed them soon after their initial collision. The name of the system refers to the striking illusion that one of the clusters looks like a bullet piercing the other. Two completely different types of observations were simultaneously made: NASA's Chandra X-ray Observatory satellite measured the x-ray emissions, while gravitational lensing studies identified the location of the mass. In the color version of Figure 2.14, the regions highlighted in red show the x-ray-emitting gaseous material. The blue regions highlight the bulk of the mass of the system. (The colors in the pictures were chosen to illustrate the separate components and are not the actual colors observed.) The red regions are made of ordinary atomic material, identified by its x-ray emission, whereas the blue regions consist of dark matter, identified by lensing.

During the initial collision of the two clusters, the atomic (red) material lost a great deal of energy and settled into a single clump at the interaction region. In contrast, the dark matter underwent essentially no interactions as it passed right through the center. Thus, the dark matter can be seen as two massive (blue) areas on either side of the central merged gas. The remarkable Bullet Cluster gives a striking confirmation of the disparate behavior of collisionless dark matter and that of ordinary atoms—exactly as predicted by the theory.

In Chapters 5 and 7, we'll see that the most likely explanation for the nature of dark matter is that it consists of some new kind of fundamental particle—not neutrons, protons, or quarks, but something entirely new. The dark matter particle feels gravity, but it does not experience the other forces of our daily experience.[11] For that reason, dark matter particles in the Bullet Cluster can move right through the middle of the cluster without stopping in the center. The gas particles slow down substantially when they hit one another; but the dark matter particles largely just keep right on going.

The Bullet Cluster (almost) puts a nail in the coffin of alternate theories of gravity. An alternative explanation to the theory of dark matter has been proposed for galactic rotation curves: Modified Newtonian Dynamics (MOND). MOND purports to eliminate the need for dark matter by instead modifying the behavior of gravity on large scales. We know from experience that Newton's laws of gravitational force work perfectly well in our everyday lives here on Earth. Most of the physics of the Solar System matches Newton's laws as well. However, even in the Solar System, the laws have to be modified to take into account the effects of Einstein's relativity. The precession of Mercury can only be explained by using General Relativity. The Global Positioning System (GPS) only works if the gravitational redshift of Einstein's theory is accounted for. Clocks keep different time here on Earth than in satellites above the Earth. The slight difference in the pull of Earth's gravity, depending on the height above Earth, is responsible for clocks recording different times. Einstein's General Relativity modifies Newton's laws of gravitation.

It's possible that, on even larger scales of galaxies, Einstein's theory is imperfect and needs to be modified further. In 1983, Mordehai Milgrom proposed

MOND to explain flat rotation curves without the existence of dark matter. In his theory, Newton's laws of gravitational force work just fine here on Earth but need some correction farther out in the Galaxy. Milgrom hypothesized that Newton's law must be modified for the extremely low accelerations at the outer reaches of the galaxies. In his model, the force of gravity changes at distances far from the center of the Galaxy.

The downside of MOND at the outset was that (unlike Einstein's relativity) it was theoretically unmotivated and aesthetically ugly—just a kluge to match the data. It's also difficult to test, because it doesn't make any new predictions on the small scales where we can make measurements. MOND requires a modification to Newton's laws that is simply tacked on at large scales, where the only observations are the rotation curves. Subsequent to the original version of MOND, more attractive variants have been put forward, such as the tensor-vector-scalar model (TeVeS) of Jacob Beckenstein. The main improvement that TeVeS offers is that it is a complete and theoretically consistent model. Unlike the earlier versions of MOND, energy and momentum are conserved, and signals do not propagate faster than the speed of light. The price paid by this theory is that many new ingredients, implying many new particles and interactions, have to be added without explanation.

In the absence of dark matter, the Bullet Cluster in Figure 2.14 is hard to explain. The dark matter explanation seems so simple: the gas undergoes a lot of collisions and gets stuck in the central region, while the dark matter interacts so weakly that it passes right through the middle. The dark matter theory can easily explain the existence of the separated blue and red regions in the Bullet Cluster. This separation of the two components is a challenge for the alternative models of MOND and TeVeS. In those models, there is only one type of matter— ordinary matter—and no simple explanation exists for the observed split into the red and blue regions. Many scientists contend that this is conclusive evidence that alternatives to the dark matter theory are dead. Yet others continue to work on these ideas based on modifications to standard gravity. The majority of the physics community believes that dark matter is the better, less contrived explanation. In addition, as discussed in the next chapter, a consensus picture of the Universe has emerged from a multitude of cosmological observations. In this picture, dark matter plays the pivotal role of dominating the mass in the Universe.

## Formation of Galaxies and Clusters

Galaxy formation provides further indirect evidence for the existence of dark matter. The web of galaxies that we live in today could not have formed without a little help from dark matter. Scientists who simulate the formation of structure in the Universe on computers have converged on a scenario known as hierarchical structure formation. Galaxy formation proceeds from the bottom up. At first, all particles were spread out (almost) uniformly.[12] Then small clumps of mass began to form. Thanks to the excess mass inside them, these clumps exerted extra gravitational pull that dragged in even more material. The first small clumps merged together to form ever-larger objects, eventually creating galaxies and clusters. Computer simulations follow this process of structure formation up to the present day. In the final results of these simulations, dark matter weaves a web of structure throughout the universe with galaxies and clusters at the nodes of the web. The resulting objects are essentially halos of different masses, consisting primarily of dark matter.

Without a dominant cold dark matter component, galaxies would never have been able to form. The dark matter accumulated first, and then pulled in the atomic material later. Dark matter's gravity created deep potential wells (regions of strong gravitational attraction) where protogalaxies began to form, and the ordinary atoms eventually fell into these wells to make the galaxies and clusters we see today.

Initially the Universe was ionized, with atoms split into positive ions and negative electrons. As long as the atoms were in charged form, they were pulled along with the photons and prevented from clumping together. Only once the ions joined with the electrons, at a time 380,000 years after the Big Bang, were the atoms freed from being swept along by photons. At that point, the atoms were able to fall into the dark matter structures that already existed. This physics is explained in detail in Chapter 3. In fact, it is likely that dark galaxies exist, with no bright stars in them, simply made of large amounts of dark matter and some gas.

When the Universe was about 200 million years old, minihalos formed. These objects, containing both dark matter and atoms, weighed a million times the mass of the Sun. At the centers of minihalos, the first stars began to form. These provided the first stellar light in the Universe and ended the epoch known as "the dark ages."[13] Early stars looked very different from today's stars. They consisted entirely of hydrogen and helium, which were the only elements that had been produced in the primordial Universe. In my work, I proposed the idea that the very earliest stars may have been "dark stars" (see Afterword: Dark Stars).[14] Although made primarily of ordinary atoms, dark stars would have been powered by dark matter and could have grown to be very big and bright. They could have become 100 or even 1 million times as massive as the Sun and up to 1 billion times as luminous. Once the dark matter fuel was exhausted, the stars would have collapsed to become smaller and hotter. In time, fusion became the power source for the stars and produced the elements of our current Universe: everything from carbon and nitrogen to lead and uranium.

The million-solar-mass dark matter minihalos that hosted the first stars merged together to make larger objects. Eventually, galaxies formed, including the Milky Way, weighing up to a trillion times as much as the Sun. These galaxies then merged to make clusters and superclusters, all predominantly consisting of dark matter.

Computer studies of galaxy formation follow this growth of structure from the smallest dark matter objects to the largest structures in the Universe. Computers can simulate the motions of billions of "particles," as they start out more or less uniformly distributed; they then form larger and larger clumps as time goes on. The "particles" in the simulation are actually dark matter clumps as large as a million times the mass of the Sun. Today's computers cannot simultaneously resolve all the physics from subatomic scales to galaxies; the best they can do is to start from these much larger mock particles. Even these require the best state-of-the-art computations run on the largest supercomputers now available. As the resolution of computers improves, astrophysicists refine their simulations and continue to learn more about galaxies and their formation.

The formation of structure depends on the type of dark matter. Current simulations assume that the dark matter is cold, that is, it is moving nonrelativistically (typically with speeds a thousandth of the speed of light). As we'll see in Chapters 5 and 7, cold dark matter candidates include supersymmetric particles, axions, and primordial black holes. Much evidence points to the validity of the cold dark matter hypothesis. The alternative hot dark matter hypothesis, where the dark matter is moving at nearly the speed of light, has been shown to be incompatible with observations of galaxies. The rapidly moving hot dark matter particles would pull apart structures as they began to form. Hence, the hot dark matter hypothesis has been discarded. Galaxy formation with cold dark matter takes place much more quickly. It produces structure in the hierarchical pattern we have been describing, where small objects merge to make ever larger ones.

To understand the big picture of the distribution of dark matter on the largest cosmological scales, we can follow a time sequence of images resulting from a simulation. The simulation shown in Figure 2.15 was run using only dark matter particles, without taking into account the effects of atomic matter.[15] Because it is the dark matter that drives the dynamics of structure formation, such simulations are thought to be quite informative. In the figure, the bright regions indicate the locations of the highest concentration of dark matter—not of stars.

The time sequence in Figure 2.15 is labeled in terms of the redshift $z$: higher values of redshift correspond to earlier times in the history of the Universe. Because the Universe is expanding, distances between objects are being stretched as time goes on. As we look backward in time, the Universe becomes more compact. The value of the redshift in each of the panels indicates how much more condensed the Universe was then than it is now. At $z = 29$, in the first panel, the Universe was 30 times more compact than it is now. At this point, the Universe was 100 million years old.

The sequence of panels in Figure 2.15 illlustrates the formation of structure from a relatively smooth beginning all the way to today's clumpy Universe. In the $z = 29$ panel (the earliest time shown in the figure), the particles were spread out almost uniformly throughout the volume of the simulation. Slight overdensities—regions with a little extra matter than average—were put in by hand. The series of figures then shows the time evolution of these dark matter regions. Small structures formed first, and then merged to make larger ones. First, sub-Earth mass objects formed, then ever larger objects, eventually leading to galaxies, and then clusters of galaxies. The end result is seen in the $z = 0$ panel of Figure 2.15, corresponding to the Universe today. The longest stringy structures are known as filaments. Galaxies and clusters are distributed at the nodes of these filaments, as indicated by the brightest spots in the figure. Most of the mass in the universe is located along these filaments.

In between the filaments are dark regions known as voids. These are the emptiest places in the Universe. They contain very little matter—roughly a tenth as much as the average diffuse regions of the Universe. However, they are not completely empty. Nowhere in the Universe is there a complete vacuum, devoid of all matter and energy.

## Dark Matter Dominates

Though dark matter does not shine, astronomers have inferred its existence from its gravitational effects on visible matter. Almost all the mass in galaxies and clusters, including our own Milky Way, is made of this unknown nonluminous material. The stars reside in a flattened disk, whereas the dark matter sweeps through a much larger spherical halo. Rotation curves of galaxies are flat: gas at great distances from the centers of galaxies is orbiting with speeds far in excess of what can be explained by the pull of stars in the

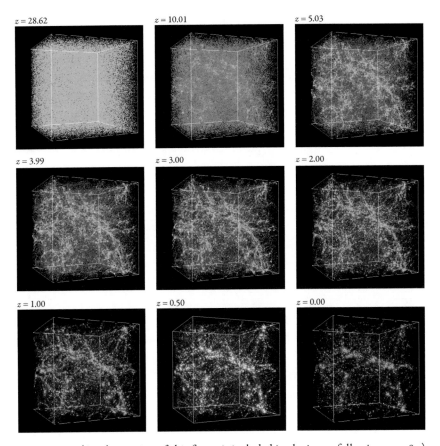

**FIGURE 2.15** (A color version of this figure is included in the insert following page 82.) Computer simulation of galaxy formation starting from 100 million years after the Big Bang ($z = 28.62$). The time sequence is labeled in terms of the redshift $z$, where higher values of $z$ correspond to earlier times in the Universe ($z = 0$ today). The bright regions in the images are actually the locations of dark matter; as the dominant matter in the Universe, it controls the formation of large-scale structure. The first small clumps of dark matter merged to form ever-larger objects, eventually creating the galaxies and other large structures we see today. Galaxies are located at the intersections of the long stringy filaments shown in the final images. Without dark matter, galaxies would never have formed and we would not exist! *Simulations were performed at the National Center for Supercomputer Applications by Andrey Kravtsov (University of Chicago) and Anatoly Klypin (New Mexico State University). Visualizations by Andrey Kravtsov.*

galaxies. A vast reservoir of dark matter must be out there to speed up these orbits. On theoretical grounds, Jeremiah (Jerry) Ostriker and James (Jim) Peebles in the 1970s argued that galactic disks would be inherently unstable unless they were surrounded by a massive spherical halo component. Their work had important impact in persuading the physics community that dark matter had to be there.

Clusters of galaxies have additional dark matter in between the galaxies. Einstein's relativity tells us that mass bends light, and this gravitational lensing shows that huge amounts of dark matter exist inside the clusters. HST has found beautiful cases of multiple, sheared images of distant stars that can only be explained by a vast dark matter reservoir bending the light. X-ray images of clusters find a large amount of hot gas in them, which can be explained only if dark matter is there to pull in the gas.

Computer simulations study the formation of galaxies and large-scale structure. Millions to billions of particles start out uniformly distributed and then clump together to form first small structures, then galaxies, and then clusters. Dark matter dominates the mass and the dynamics on all scales of structure, from galaxies to clusters to these large filaments.

Yet so far, all that we know about dark matter is that it feels the force of gravity. We still have no clue as to its identity. Other than gravitationally, does it interact at all with ordinary matter? In later chapters, we'll see that astrophysicists believe that it does and that these interactions in detectors provide tools to discern its nature. By studying the big picture of the Universe, in the next chapter we will see the role dark matter plays in shaping the overall cosmological evolution of the Universe.

# THREE

# The Big Picture of the Universe
## Einstein and the Big Bang

The combination of Albert Einstein's theoretical insights in 1915 together with the data from this century have led to remarkable developments in cosmology. Over the past 20 years, the cosmological data have been pouring in, and the amount we have learned about the big picture of the Universe has been breathtaking. This chapter discusses the theoretical ideas and observations behind modern cosmology and describes how dark matter fits into the big picture of the Universe.

## Geometry of the Universe
### Einstein and Warped Geometry

A scientific approach to the question of the shape and basic structure of our Universe dates back to brilliant insights of the early twentieth century. Albert Einstein published his theory of General Relativity in 1915, based on simple but counterintuitive ideas with deep and far-reaching consequences. His equations relate mass to the curvature of spacetime. The application of these concepts to the Universe as a whole gave rise to an overall picture of its structure.

According to General Relativity, the heavier a body is, the more it *warps* the spacetime around it. We've already seen an example of this phenomenon in Chapter 2, where we discussed gravitational lensing as a tool to search for dark matter. When light passes near any mass—such as the Sun, the Earth, or dark matter—its path is bent toward the massive object. The bending of light by gravity was first tested in 1919 by observations of distant stars beyond the Sun. The starlight traveled along a curved path around the Sun to telescopes on Earth.

We can carry this picture one step further. Instead of viewing the light path as being curved as it passes near the Sun, there's an alternative perspective: the underlying space around the Sun can be thought of as curved, like the surface of a basketball. The light travels in a "straight line" along this curved background.

An analogy would be an airplane traveling from New York to California. The most direct route would be for the plane to follow the curvature of the Earth. The airplane's path can be thought of as a *geodesic*—the path of shortest distance between two points—on a spherical surface above the Earth. We have two different perspectives for the same reality: curved paths in ordinary geometry or straight lines in curved backgrounds. This second approach is mathematically simpler as well as more fundamental. Any massive body like the Sun warps the space around it.

This connection between mass and geometry is the essence of Einstein's General Relativity and carries over to the Universe as a whole. Another analogy would be a waterbed with a bowling ball placed on top of it. The waterbed would sag under the weight of the bowling ball, leading to a warped surface. Any other much lighter object placed on the waterbed would roll down along a "straight line" toward the bowling ball. Similarly, the total mass/energy content of the Universe determines its curvature and its geometry. Einstein's famous equation $E = mc^2$ implies that the mass $m$ of a body is a component of its energy $E$. Thus mass and energy can be converted into one another. A body's total energy is given by its mass plus its kinetic (movement) energy. It is the sum total of the mass and energy in the Universe that distorts its spacetime and determines its geometry.

By combining principles of cosmological symmetry with these concepts from Einstein's relativity, scientists in the 1920s postulated an entirely new vision for the Universe. The rest of this chapter illustrates these astounding concepts, their history, and their observational confirmation.

## Hubble Expansion

In the 1920s and 1930s, Alexander Friedmann, Georges Lemaître (a Belgian priest), Howard Robertson, and Arthur Walker created the modern framework for cosmology. They proposed two cosmological principles for the Universe: *homogeneity* and *isotropy*. They suggested that the Universe is homogeneous (looks the same at every point) and isotropic (looks the same in every direction). These principles apply to the Universe on the largest scales—that is, when we average over the galaxies and other structures. On small scales the individual objects clearly can be distinguished from one another (for example, here on Earth, where no two people are the same). The key point of these two concepts is that there is no central point or special direction in the Uni-

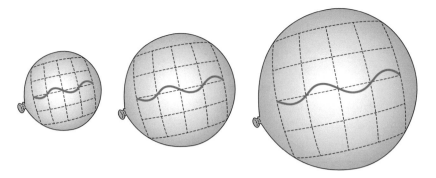

FIGURE 3.1 Wavelengths of light stretch as the Universe expands, as shown by analogy with an expanding balloon.

verse. Observations of the Cosmos made here on Earth would look the same, on the average, as observations made by an astronomer in a distant galaxy. Mathematically, these notions of homogeneity and isotropy correspond to precisely defined symmetries. Friedmann, Lemaître, Robertson, and Walker then applied these two cosmological principles to Einstein's equations of relativity to develop the equations that describe the evolution of our Universe as a whole.

In the 1920s, in addition to homogeneity and isotropy, Einstein wanted to add a third symmetry. He postulated that the Universe is static and immobile: that is, it looks the same (on average) at all times. For the mathematics to be consistent with such a time-symmetric Universe, he was forced to add another term—known as a cosmological constant—to his equations of General Relativity. He preferred the idea of a static Universe, because physicists think of symmetries as elegant, simple, and beautiful. But in this case, Einstein turned out to be just plain wrong.

In 1929 Edwin Hubble, using the Hooker Telescope at Mount Wilson Observatory in the San Gabriel Mountains above Pasadena, California, made an astonishing discovery. He observed light from galaxies at various distances away from Earth. Atoms emit light at characteristic wavelengths, producing "atomic lines" that are unique to the type of atom. Hubble found that the wavelength of atomic light emitted by galaxies had been redshifted, or stretched, by the time it reached the telescopes. The implication of these observations was enormous. To explain the data, Hubble came to the conclusion that the Universe must be expanding. Figure 3.1 illustrates this concept of the stretching of light waves by the expansion of space.

As a consequence of this *Hubble expansion,* galaxies all appear to be moving apart from one another. Hubble found that, the farther a galaxy is away from Earth, the faster it appears to be receding from us. If we drew dots on a balloon and inflated it, the same would happen: the farther apart the dots, the faster they would move apart. He encoded this pattern in a simple formula now known as Hubble's law: $v = Hd$, where $v$ is the speed, $d$ is the distance from us to the galaxy, and $H$ is the rate of the Universe's expansion, known as the Hubble constant. It was Hubble's publication of data confirming this law that convinced the scientific community that our Universe is expanding.

Hubble's data were rough, and his estimate of the expansion rate of the Universe was almost 10 times too fast. Much later, astronomers were able to determine the speed of the Universe's expansion more accurately. In the past decade several groups have found a Hubble constant of approximately $H = 70$ kilometers per second per megaparsec (though they still disagree[1] by up to 10%). These strange units correspond to what observers measure. Galactic speeds are measured in kilometers per second. A megaparsec is roughly 3 million light-years, a typical distance between two large galaxies. Because the Hubble constant is a measure of the stretching of space, cosmologists can use it to extrapolate backward to determine how old the Universe is. The result of this extrapolation implies that the Universe is almost 14 billion years old.[2]

Hubble's measurements unequivocally showed that the Universe is expanding (Figure 3.2). Einstein was forced to abandon what seemed to him an aesthetically more perfect idea of an unchanging Universe. The growing distance between galaxies is not a consequence of their independent movement; instead, it is simply a consequence of the expansion of space. Einstein had originally introduced a type of vacuum energy called a cosmological constant into his equations to make the Universe static. After Hubble proved that the Universe is expanding, Einstein rejected his own theory of the cosmological constant as his "biggest blunder." In the past 15 years, however, the cosmological constant has made a surprising comeback as a possible explanation for the acceleration of the Universe. We'll return to these concepts when we discuss dark energy in Chapter 9.

Although on average galaxies are moving apart from one another on very large scales, on smaller scales, gravity can dominate over the expansion. The Milky Way and its nearest big neighboring galaxy, Andromeda, also known as M31, are being pulled together by gravity. In roughly 4.5 billion years, these galaxies will merge. Similarly, galaxies themselves are self-gravitating objects that have "decoupled" from the Hubble expansion. Our Milky Way, for example, is so massive that as residents inside it, we are completely immune from the effects of the expansion. It is only on large scales of the Universe that the expansion dominates.

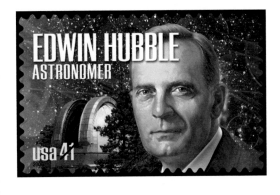

**FIGURE 3.2** In 2008 the U.S. Postal Service released a 41-cent stamp honoring Hubble with the following citation: "Often called a 'pioneer of the distant stars,' astronomer Edwin Hubble (1889–1953) played a pivotal role in deciphering the vast and complex nature of the universe. His meticulous studies of spiral nebulae proved the existence of galaxies other than our own Milky Way. Had he not died suddenly in 1953, Hubble would have won that year's Nobel Prize in Physics." *USPS / Victor Stabin.*

An analogy that is often used for the expanding Universe is the "raisin bread model" of the Universe. Imagine a piece of raisin bread that expands when you put it in the oven (I'm not much of a cook, but I'm told the dough rises). The dough provides an analogy to the Universe as a whole, and the raisins in the bread are like galaxies. The raisins are not in any sense running away from one another. Instead, they separate as time goes on because the bread is expanding in between them. Similarly, the galaxies in the Universe are not all racing apart from one another; rather the space in between them is growing larger because of the expansion of the Universe.

There is an important difference between the raisin bread model and the real Universe. Whereas the raisin bread has a central point in the middle, the Universe does not. The cosmological principle of homogeneity implies that there is no special point in the Universe—no center. From our perspective, we see other galaxies moving away from us due to the expansion; yet this does not imply that we are at a central point in the Universe. If we were to transport ourselves to another galaxy, we would see exactly the same thing: again all galaxies would be moving away from us. From this new vantage point, the Milky Way Galaxy would be retreating just like any other galaxy. Similarly, from the perspective of any one of the raisins in the bread, the other raisins are moving farther and farther apart.

Hubble made a second discovery that completely altered the prevailing worldview about the Universe. Until the early twentieth century, scientists believed that all the stars in the night sky belonged to a single galaxy, our own Milky Way. The idea that the Universe could be large enough to extend beyond our Galaxy seemed incomprehensible. In the late 1700s, as Charles Messier was using his telescopes to search for comets, he was irritated by the many cloudy objects, or nebulae, that obscured his observations. He published a catalog of about 100 "nebulae," which, to this day, are still known as Messier objects. In fact, most of these nebulae are galaxies. He had unwittingly made the first catalog of galaxies, though he remained completely unaware of the type of object he was looking at. These objects are still classified by the designations M1 to M110, with the "M" standing for Messier.

A hundred years ago, scientists were still unaware that anything existed outside of our own Galaxy. It was Hubble's observations in the 1920s that showed that some of the starlight observed by Messier in fact comes from beyond the Milky Way. Hubble measured the distances to bright objects and concluded that they were unequivocally too far away to lie inside the Milky Way. He was the first to prove that galaxies existed besides our own, and he dispelled the prevailing view that the Milky Way was the only structure in the Universe. Our Universe is so much vaster than was appreciated even a century ago!

Testing the principle that the Universe is homogeneous required observers to look at the Universe out to great distances, beyond our own Galaxy and beyond nearby clusters of galaxies. These measurements turned out to be surprisingly elusive. The goal was to make a survey of galaxies on large enough scales to probe the average properties of the Universe. In the 1980s, Margaret Geller and John Huchra at the Harvard / Smithsonian Observatory with the Center for Astrophysics Survey expected to find evidence for homogeneity. Instead, they discovered the Great Wall of galaxies, a structure that stretches 100 million light-years across the sky. Finally, in 2002, the 2dF Galaxy Redshift Survey using the Anglo-Australian Telescope in eastern Australia looked out far enough into the Universe. By studying more than 200,000 galaxies out to billions of light-years away from Earth, these observations for the first time verified that on the largest scales, the Universe does look homogeneous.

### Shape of the Universe

When Friedmann, Lemaître, Robertson, and Walker applied the two cosmological principles of homogeneity and isotropy to Einstein's equations in the 1920s, a theoretical framework for describing our Universe as a whole emerged. As a consequence of these symmetries, these early cosmologists deduced that there are three possible shapes, or geometries, for the Universe, known as flat, spherical, and hyberboloid (saddle shaped). Figure 3.3 illustrates two-dimensional analogues of these shapes; three dimensions of space are impossible to plot on a piece of paper.

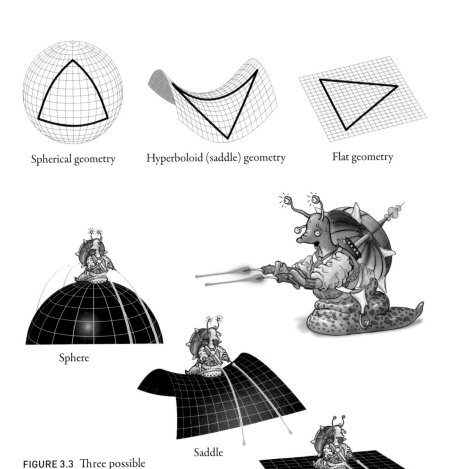

Spherical geometry     Hyperboloid (saddle) geometry     Flat geometry

Sphere

Saddle

**FIGURE 3.3** Three possible geometries for the Universe, circa 1930. *(Bottom) NASA/EMAP Science Team/B. Griswold.*

Flat

The three different possible geometries for the Universe correspond to different *curvatures* as illustrated in the figure. The notion of a *flat* geometry does not imply that the Universe is two dimensional (Figure 3.4). Clearly there are at least the three dimensions of our daily experience (if modern string theories are correct, there are six more). The word "flat" refers instead to the fact that the Universe has no curvature—no weird geometry. When the snail sitting on the flat surface in Figure 3.3 sends out two parallel light beams with his phasers, the beams will continue in straight lines all the way out to infinity without ever intersecting. In a flat geometry, the shortest distance between two points is a straight line—the usual situation in our daily experience.

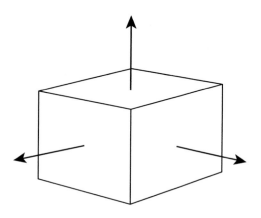

**FIGURE 3.4** A Universe with a flat geometry is not two dimensional. Instead, it goes out to infinity in all three directions (as indicated by the arrows). "Flat" refers to the fact that the Universe has no curvature.

In contrast, a spherical geometry can be envisioned by imagining a creature sitting on top of a sphere (like a basketball). The snail on top of the ball in Figure 3.3 could send out two light rays in what seem to be parallel directions, but because the rays follow the curvature of the sphere, they would eventually meet again when they returned to the North Pole. If nothing interfered, his phaser shots would eventually hit him from behind. The shortest distance between two points—the geodesic—would not be a straight line; instead, it would follow the spherical curvature of the basketball. The third possibility for the geometry of the Universe is *hyperboloid*, that is, saddle-shaped. In this case, two parallel light rays would diverge from one another and would never meet again.

In the case of a Universe with spherical geometry, one concept that puzzles people greatly is the question: what is inside the sphere? In fact, there is nothing in the interior. The Universe exists only on the surface of the sphere. The surface is a pictorial representation of all possible geodesic paths connecting material objects, such as the paths of light rays sent out by the snail described above. In this case, arcs on the sphere represent the shortest distance between two points.

Another distinction between the three possible geometries is the difference in the sum of the three angles in a triangle, as illustrated in the top part of Figure 3.3. In a flat geometry, the angles in a triangle always add to 180 degrees (as students learn in high-school geometry classes). In a spherical geometry, the angles add to more than 180 degrees; the exact number depends on the sphere. In a hyperboloid geometry, the three angles in a triangle add up to less than 180 degrees.

The three different geometries correspond to different numerical values of the curvature of space. In Einstein's equations, a Universe with flat geometry has zero curvature; a spherical Universe would have positive curvature; and a hyperboloid Universe would have negative curvature. Mathematically, it's much easier to work with lines on curved surfaces than with curved lines in flat space: the concept of curvature provides an alternate description that more concisely accounts for the effects of mass and energy in the Universe. According to Einstein's relativity, mass and energy warp spacetime and give it one of these three possible shapes.

Time can be thought of as a fourth dimension. Time appears to always move forward rather than backward. Although the origin of time is a very interesting unsolved problem, this book postulates time's forward motion without seeking an explanation.

The three geometries correspond to very different possibilities for the future evolution of the Universe. A spherical Universe would expand to a maximum size and then start to recollapse. One can visualize blowing up a balloon and then letting the air back out. The ultimate endpoint would be a "Big Crunch." Such a spherical Universe would be finite in spatial extent and only exist for a finite amount of time. The other two cases, the flat and hyperboloid Universes, do not contain enough matter and energy to provide the gravitational attraction that leads to recollapse. Instead, flat and hyperboloid Universes would expand forever, leading to an ever colder and emptier Universe, ultimately approaching a "Big Chill." An interesting variant exists, though, for the spherical Universe. It is possible that, after its recollapse, it bounces and starts expanding again. Such a Universe would follow a cycle of a Big Bang followed by a Big Crunch, and then back to the next Big Bang. Chapter 9 further discusses various possibilities for the final fate of the Universe.

## Cosmic Hints

How odd that the scientists of the past century who created the field of cosmology died without knowing whether their ideas were right. Theoretical physicists always have this uneasy feeling that they may be working on science fiction. No matter how brilliant and creative an idea is, it just may not turn out to be the way the world works. Getting the right answer also requires a lot of luck. Most good ideas turn out to be wrong when new data show that they just don't match our Universe. The creators of the modern framework for cosmology died before their ideas could be confirmed. In fact, they were completely correct.

Since the 1920s the determination of the curvature, mass, and energy density of the Universe defined a deep unanswered quest for cosmologists. Of the three possible geometries for the Universe, which one is right? Theorists argued for a flat Universe. On theoretical grounds, it seemed to be the only case that made any sense. Only a flat Universe could have lasted long enough to allow galaxies to form and life to exist. In the case of a spherical or hyperboloid geometry, the Universe should only have lasted for an instant after the Big Bang. A spherical Universe would logically have recollapsed instantaneously into a Big Crunch, whereas a hyperboloid Universe would have rapidly expanded into a cold uninhabitable Big Chill. Yet we exist. That we avoided either of these fates implies that our Universe must have zero curvature, or nearly so. Only with a flat geometry can the Universe grow old enough to create the conditions for life to exist.[3]

In contrast, astronomers simply could not find enough matter in the Universe to be consistent with a flat geometry. Neta Bahcall was one of the leaders of the groups that made these observations. In the 1980s and 1990s, she examined rich clusters, which contain hundreds of galaxies. As a test case, she considered the extreme assumption that these massive clusters were representative of the Universe as a whole. She knew this assumption would produce an overestimate of the total mass content of the Universe. Still, she found that the total amount of mass in the Universe could add up to at most a third of what is required for the Universe to be flat. I remember Neta giving talks at conferences, and theorists persuading themselves she was wrong. They believed that Neta had simply missed some of the matter in the Universe. Elegantly dressed, she stressed repeatedly that the matter content of the Universe plateaued at a level too low to produce a flat Universe. Yet in most of the physics community, the theoretical prejudice for a flat Universe made entirely of matter persisted. Eventually it became clear that, as far as the mass content is concerned, Neta was right. This conundrum, the apparent inconsistency of the geometry and the mass of the Universe, is the subject of the rest of this chapter.

## Cosmic Microwave Background

This question of the geometry of the Universe has been a holy grail for cosmologists since the 1920s. At the time I started working as a graduate student in the 1980s, no one thought that it would be possible to differentiate observationally between these three possible geometries any time soon. Astronomers tried to add up all the known pieces of the Universe to determine its entire content but were always aware that they could be missing some important constituent. Then at the turn of the millennium, observations of the cosmic microwave background (CMB) finally provided some answers.

Today we are bathed in the faint glow of microwave radiation left over from the hot early era of the Universe after the Big Bang. This CMB gives us a snapshot of the primeval Universe and is one of our best probes of cosmology. Discovered in the 1960s, the CMB provided one of the first compelling pieces of observational evidence for the validity of the Hot Big Bang model. Although Hubble had proved that the Universe is expanding, still a variety of possibilities remained for its nature at early times. Did it start out hot or cold? Scientists argued in favor of a number of models, including a Hot Big Bang and a Cold Big Bang. As we'll see, observations of the CMB helped to settle this argument. Without a hot early phase, the CMB would not have come to exist. Only the fierce heat of the early universe from the Hot Big Bang can explain the existence of this remnant thermal radiation.

Here the word "radiation" refers to electromagnetic radiation, or particles of light—rather than to the radiation from nuclear bombs or nuclear reactors. The early Universe was a primordial soup of fundamental particles smashing into one another. These included particles of light (photons). Though at first photons were constrained to short paths by their constant interactions with other particles, in time they were liberated and have been able to traverse the Universe unimpeded en route to our telescopes. These photons constitute today's CMB.

In its infancy, the Universe was opaque. Instead of neutral atoms, it contained positively charged ions separated from negatively charged electrons. Electrons roamed freely, unattached to hydrogen or helium from the Big Bang. Photons couldn't travel even a millimeter without running into these electrons and being deflected from their paths. If we could somehow travel back to that early epoch without being fried by the heat, our surroundings would be pitch black. Light from other objects could never reach us without scattering off of electrons en route. We might see occasional flashes of light as photon-electron interactions took place nearby.

A drastic change happened once the Universe cooled to a temperature of 3,000 Kelvins (K), approximately 380,000 years after the Big Bang.[4] A transition took place known as *recombination* (Figure 3.5). The free electrons that had previously scattered the photons disappeared as they became bound into atoms. Reaction rates depend on temperature, and at this point, protons (denoted by $p^+$) and electrons ($e^-$) combined to create hydrogen plus photons:

$$e^- + p^+ \rightarrow H + \gamma.$$

Here, $\gamma$ indicates photons coming out of the reaction. The Universe had cooled to the point where hydrogen could no longer be dissociated, or broken apart,

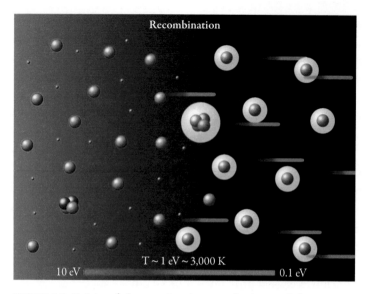

**FIGURE 3.5** Origin of the cosmic microwave background (time moves from left to right). Photons (light particles) scattered off of electrons until recombination at a temperature of 3,000 K; after that the free electrons combined with protons to form neutral hydrogen atoms. From that time the photons (indicated as streaks) travel to us unimpeded and today are detected as microwaves. *Professor William Kinney of the University at Buffalo.*

into its constituent protons and electrons. As hydrogen became stable, the Universe changed from ionized to electrically neutral. Hydrogen atoms are very ineffective at scattering light. Photons interact only with electrically charged particles, like electrons, and not with hydrogen atoms. The average light particle in the Universe has never interacted again since its last scattering with free electrons at 3,000 K. These photons travel through space and time all the way to our modern telescopes, and we detect them today as the CMB.

Today we can detect these photons arriving from the time of *last scattering* more than 13 billion years ago. Figure 3.6 illustrates this concept by analogy to a cloudy day. When we look out at the sky, we can only see as far as the clouds. This is the point where light traveling through the atmosphere was last scattered. Light from behind the clouds is obscured because of its interactions with water droplets. Similarly, we can only look back into the early universe as far as the last interactions of photons in the early Universe. The CMB we observe comes to us from this surface of last scattering.

Since then the Universe has continued to expand and cool. Today it is about 1,000 times as large as it was at last scattering. Because of this expan-

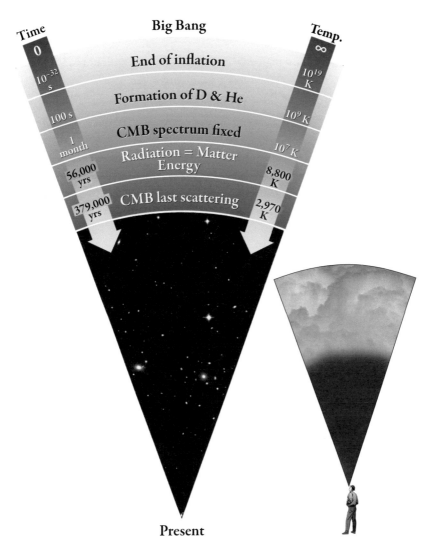

FIGURE 3.6 Photons travel to us unimpeded from the surface of last scattering, when they last interacted with electrons. This surface is analogous to a layer of clouds. In both cases, we can see only as far out as the surface corresponding to the last time light scattered. *NASA.*

sion, the CMB photons have also cooled by about a factor of 1,000. A similar phenomenon takes place with hot gas confined in a box: if you slowly pull out one wall of the box and allow the gas to expand, the temperature of the gas drops by an amount precisely determined by the increased volume of the box. At the time of last scattering, the temperature of the universe was roughly 3,000 K. Today the temperature of the CMB photons is measured to be 2.76 K.

Because of the expansion of the Universe, the wavelengths of the CMB photons have stretched since last scattering (similar to Figure 3.1). Today their wavelengths are roughly a few centimeters, characteristic of those of microwaves. Hence the word "microwave" in "cosmic microwave background." The word "background'" refers to the fact that these photons can be found everywhere in the universe. They even cause static on television screens. The CMB photons have slightly shorter wavelengths than the radiation in microwave ovens.[5]

## History of the CMB

Three scientists, Ralph Alpher, George Gamow, and Robert Herman, first predicted the existence of the CMB in 1948. Yet for almost two decades it remained only a theoretical concept. In 1964, two astronomers at Bell Labs in New Jersey, Arno Penzias and Robert Wilson, made an accidental discovery. In the process of adapting a large antenna (originally developed for tracking communications satellites) for radio astronomy, they found troublesome noise signals. At first they attributed these to various instrumental problems. They even climbed into the antenna to scrape out pigeon excrement. In the meantime Princeton physicists Robert Dicke, James Peebles, and David Wilkinson had been building a radio antenna with the explicit purpose of looking for the CMB. After communicating with the Bell Labs astronomers, the Princeton group immediately realized that Penzias and Wilson's noise was actually the first discovery of the CMB. The Princeton group quickly confirmed the result with its own experiment. To quote Dicke speaking to his collaborators after a phone call with Bell Labs, "Boys, we've been scooped!" Penzias and Wilson were awarded the Nobel Prize in 1978.

At the time the CMB was discovered, the Hot Big Bang model still had several competitors, among them the Cold Big Bang model and the Steady State Universe. In the Cold Big Bang picture, the Universe expanded outward from a dense initial state that was cold instead of hot. The Steady State model relied on matter that would mysteriously appear throughout the Universe at all times; the Universe expanded without beginning and without end. Neither of these models ever experienced a hot phase. The Steady State model was intrinsically beautiful in the sense that it satisfied a perfect cosmological principle of homogeneity and isotropy in both space and time.[6]

In the 1990s, NASA's Cosmic Background Explorer (COBE) satellite tested these alternative models. If the Hot Big Bang were right, the CMB photons should have a *blackbody spectrum*. This is a characteristic signature of the Hot Big Bang and would prove that these photons came from a hot early

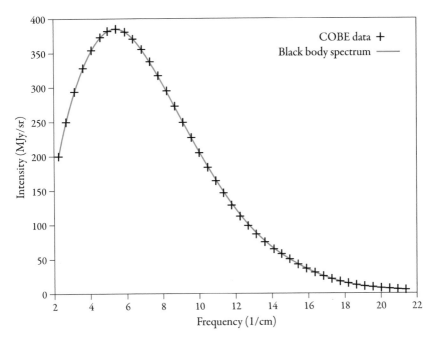

**FIGURE 3.7** Blackbody spectrum of the cosmic microwave background. The plot shows the intensity of light versus the frequency (higher frequency corresponds to shorter wavelength). Theoretical predictions (solid line) perfectly match data (crosses) taken in 1990 by the Cosmic Background Explorer satellite. *NASA/COBE team.*

phase of the Universe. According to the Hot Big Bang, the early Universe was a hot dense primordial soup of particles constantly running into one another. Because of the rapid particle interactions, scientists predicted that early photons should acquire the characteristic blackbody shape shown in Figure 3.7, determined by the temperature. As mentioned above, today the CMB photons have a temperature of 2.76 K, corresponding to a peak wavelength of a few centimeters—characteristic of microwave radiation. The Hot Big Bang predicts a precise shape for the falloff in intensity on either side of the peak.[7]

Many hot objects, including stovetops and hot plates, emit photons with an intensity that follows the same blackbody shape. At 200 degrees F, a hot plate emits photons with a peak wavelength of 4 microns (a micron is 1 millionth of a meter). The plate will appear to be red hot, because it also radiates some photons with red wavelengths, below the peak emission. Another example of a blackbody emitter is the Sun, with a peak in the yellow/green part of the spectrum. Our eyes have adapted to our proximity to the Sun; our vision is sensitive to wavelengths of visible light that exactly match the peak wavelengths of solar radiation. The Sun also creates some ultraviolet light at higher energies than the peak. Though we can't see these ultraviolet rays, they still impinge on us and cause sunburn. If our eyes were sensitive to light with wavelengths different from the visible (the peak blackbody emission from the Sun), the world would look very different. In the movie *Predator,* for example, the creature sees the movement of humans even in darkness because of their infrared radiation. Similarly, military operations use infrared detectors for night vision, surveillance, and hitting targets. Telescopes capable of detecting many different wavelengths beyond the visible have opened new windows on the sky.

When the CMB was first studied in the 1960s and 1970s, the spectrum appeared to have the wrong shape. The expected turnover after the peak didn't seem to be present in the data. Scientists speculated about the cause of the discrepancy, with ideas ranging from stellar contamination to alternatives to the Hot Big Bang. I myself wrote a paper exploring a model of galaxy formation driven by explosions as an origin of the excess.

Then in 1989, NASA launched COBE. Within the first 10 minutes of data acquisition, COBE measured the CMB to have the most perfect blackbody spectrum ever measured—better than any here on Earth (see the data in Figure 3.7). These 10 minutes of data were enough to kill all competing models to the Hot Big Bang. Only a hot early phase could have produced a blackbody spectrum. It might be possible to imagine a scenario for generating radiation at one single wavelength in a cold Universe. Yet producing a perfect blackbody curve would be nearly impossible. The only plausible explanation is that the Universe was once hot and contained rapidly interacting photons. The CMB is a pillar of the Hot Big Bang model.

The CMB has given us other important pieces of information about the Cosmos. Among these is a test of the cosmological principle of isotropy— that on average, the universe looks the same in every direction. Cosmological isotropy has been verified by the nearly perfect uniformity of the microwave background. The same microwave photons come from every direction in the sky with the same 2.76 K temperature—or almost, as we shall see!

Anisotropies in the CMB: Discovery of the Geometry
and Total Content of the Universe

The CMB coming to us from the surface of last scattering is nearly the same in every direction we look—but not quite. The photons have traveled to us from 380,000 years after the Big Bang to the current epoch. Today they are characterized by a temperature of 2.76 K. Yet photons at different locations have slight deviations from this temperature. These small differences can be measured by telescopes today and reveal tremendous amounts of information about the Universe. Cosmologists use these small deviations from isotropy to measure the geometry and total content of the Universe.

The slight temperature variations in different directions in the sky are fingerprints of galaxy formation. In the gravitational picture of structure formation, the process started with density fluctuations produced during an early inflationary epoch. (*Density fluctuations* are variations in the mass density—the amount of mass in some unit of volume—from one place to the next.) *Inflation* is a period of exponential expansion just after the Big Bang. As a result of inflation, the amount of matter in neighboring regions of the Universe differed by a small amount.

Though the variations in mass from one region to the next were initially tiny, they grew with time. Any overdense patch of the Universe, containing more matter than its surroundings, exerted extra gravitational pull on its neighborhood. This patch attracted more matter and became even heavier. Eventually, discernable structures formed. The first ones created were probably even smaller than Earth. Then, according to the hierarchical picture of galaxy formation, they merged together to make ever larger structures. Eventually this merger process led to the formation of galaxies and clusters of galaxies.

In contrast, regions that started out underdense compared to their surroundings lost more and more mass to their neighbors. Today these regions are sparsely populated and constitute *voids*. A typical void is about five times less dense than the average Universe, whereas galaxies are a million times more dense. Voids really are quite empty.

The density fluctuations in the early stages of galaxy formation left an imprint on the CMB. Prior to last scattering, the frequent interactions of photons with electrons caused the photons and the atomic matter to move as a single fluid. As matter fell into overdense regions of the Universe, it pulled the photons along as well. The photons in these regions became slightly hotter than their surroundings. As a result, small variations in photon temperature were created. The temperature deviations of the photons from one place to another are known as *anisotropies*. Oscillations in the atom/photon fluid were permanently imprinted at the time of last scattering, and we can measure them today. Detailed studies of the slight deviations of the temperature of the microwave photons from one part of the sky to another have given rise to what is known as the precision era of cosmology.

The temperature anisotropies are tiny, and thus they are hard to measure.[8] The basic approach is to scan the sky and see whether the CMB temperature differs slightly from one direction to another. Astronomers built a variety of

**FIGURE 3.8** John Mather and George Smoot won the 2006 Nobel Prize "for their discovery of the blackbody form and anisotropy of the cosmic microwave background radiation." *(Left) NASA / Bill Ingalls; (right) photo by Roy Kaltschmidt, Berkeley Lab.*

instruments on different platforms. Some were ground based, some were balloon borne, and some were launched aboard rockets. Many generations of telescopes, sensitive to increasingly small temperature variations, failed to find anything. The astrophysics community was getting worried after decades of searches failed to find any hints of discovery. Year after year I would meet CMB experimentalists at conferences as they presented ever-better data but still found no anisotropies. I remember chatting in a jacuzzi at a winter conference with graduate students and a reporter from *US News & World Report* about the implications of the lack of observed signal. Could it mean, we wondered, that there was something seriously wrong with the Hot Big Bang?

Finally, in 1991 COBE, the same satellite that also proved that the CMB has a blackbody spectrum, found interesting results. COBE made measurements of the sky in directions that were 10 degrees apart, and found small temperature differences at the level of 1 part in 100,000. COBE was the first to discover temperature anisotropies.

This discovery proved that the gravitational instability picture of galaxy formation is correct: small density fluctuations grow to form structures. It further cemented the Hot Big Bang as the correct model of the evolution of the Universe. At the University of Michigan, I convened a meeting of the faculty to tell them about this important result. In 2006, John Mather[9] and George Smoot were awarded the Nobel Prize in Physics "for their discovery of the blackbody form and anisotropy of the cosmic microwave background radiation" (Figure 3.8).

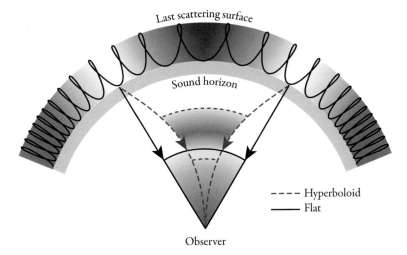

**FIGURE 3.9** The sound horizon at the surface of last scattering serves as a meter stick. In a flat geometry, light rays travel in straight (solid) lines. In a hyperboloid (open) geometry, light travels along curved (dotted) lines. Measuring the angle between these lines determines the geometry of the Universe. *From Hu, W., N. Sugiyama, and J. Silk. 1997. "The Physics of Microwave Background Anisotropies." Nature 386: 37.*

### Geometry Revealed

A decade after their initial discovery by COBE, temperature anisotropies in the CMB played an even more important role: they tested the geometry of the Universe.

Acoustic oscillations in the atom/photon fluid arising from early density fluctuations are imprinted at last scattering. The length scale of these oscillations is determined by the *sound horizon,* the distance that sound could have traveled in the Universe by the time of last scattering. This distance is well known and can be used as the equivalent of a ruler, or meter stick, in studies of the CMB.

Figure 3.9 illustrates the trick used to measure the geometry of the Universe. From either end of the sound horizon, light travels along one of the paths (indicated by arrows) to an observer on Earth. The choice of path taken by the light depends on the curvature of the Universe. If the geometry were flat, then the light would follow a straight line (solid lines in the picture). If the geometry were hyperboloid (saddle shaped), then the light would bend inward and travel along the dotted lines. If the geometry were spherical, the light rays would bow outward. The angle between the two light paths (coming from either end of the sound horizon) depends on the geometry of the Universe.

For a flat geometry, the angle would be 1 degree (shown in Figure 3.9 as the angle between the two solid lines). For a spherical geometry, the angle would be larger; for a hyperboloid geometry, the angle would be smaller than 1 degree (shown in Figure 3.9 as the angle between the two dotted lines).

A large anisotropy, known as a Doppler peak, should appear in CMB data on an angular scale that reflects the shape of the Universe. If the universe were flat, measurements of the CMB temperature that compare two regions in the sky that are separated by an angle of 1 degree should register the largest anisotropy. For the other two geometries, the peak should be at slightly different angular scales (larger than 1 degree for a spherical Universe and smaller than 1 degree for a hyperboloid Universe). In the 1990s, COBE was the first to measure anisotropy at all. However, its measurements were on a 10-degree angular scale, much too large to differentiate among the possibilities. COBE didn't have the angular resolution to distinguish among the possible geometries.

Dozens of groups competed to be the first to find the Doppler peak that would divulge the shape of the Universe. Graca Rocha of the California Institute of Technology's Jet Propulsion Laboratory wrote an early paper with collaborators that carefully combined results from a variety of experiments to argue that the Doppler peak had been found.[10] Then in 1999 Amber Miller, for her PhD thesis with advisor Lyman Page, together with collaborators at Princeton University and the University of Pennsylvania, used the Mobile Anisotropy Telescope in Chile to study the CMB. Known as the MAT/ TOCO experiment, this group for the first time succeeded in identifying the Doppler peak with a single instrument.[11] They located the peak at 1 degree, consistent with a flat geometry for the Universe! The discovery of the Doppler peak answered the almost century-old question about the geometry of the Universe and is among the greatest achievements in all of science.

Soon after, the peak was also seen in the data of two other experiments: the Balloon Observations of Millimetric Extragalactic Radiation and Geophysics (BOOMERANG) and the Degree Angular Scale Interferometer (DASI), both located in Antarctica. Later, the WMAP and Planck satellites studied the CMB with much higher resolution and located the peak with greater accuracy. We'll examine these observations in more detail, as the amount of information gleaned about the Universe is tremendous.

The BOOMERANG experiment was launched in 1998 near the South Pole. It used a telescope suspended from a balloon that circumnavigated the Pole for 10.5 days at an altitude of 120,000 feet. Figure 3.10 shows the path that BOOMERANG took, and Figure 3.11 shows the balloon as it was about to be launched.

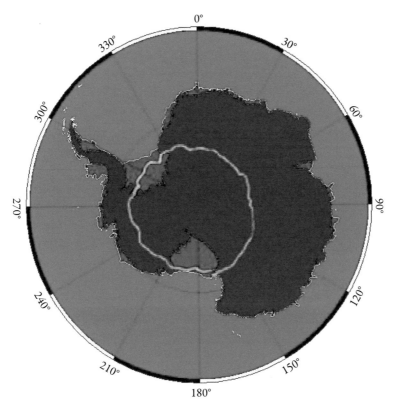

FIGURE 3.10 (A color version of this figure is included in the insert following page 82.) The path of the BOOMERANG satellite as it circumnavigated the South Pole. *The Boomerang Collaboration.*

Figure 3.11 also illustrates the results of the BOOMERANG group. In the background of the figure is a computer-reconstructed image of the patterns in the microwave sky. If we could take pictures with the equivalent of a 35mm camera in microwave wavelengths, this is the image of the Universe we would obtain. The hot spots, characterized by slightly hotter microwave temperatures, are dark, whereas the cold spots are light. Although it isn't easy to tell without a proper statistical analysis, the typical hot spots or cold spots are roughly an angle of 1 degree in size, corresponding to a Doppler peak on an angular scale of 1 degree (Figure 3.12 illustrates this peak—but the data in the figure are from much later, in 2013, and the figure is further discussed below). These patterns of structure in the microwave background revealed by the BOOMERANG data on scales of 1 degree were exactly what had been predicted for a flat geometry of the Universe.[12]

FIGURE 3.11 (A color version of this figure is included in the insert following page 82.) The BOOMERANG experiment about to be launched at the South Pole. In the background is a computer reconstruction of the microwave images it saw. From the sizes of the dark blue hot spots, scientists deciphered the shape and curvature of the Universe. *The Boomerang Collaboration.*

The leaders of the BOOMERANG experiment were Paolo de Bernardis from the University of Rome and Andrew Lange from the California Institute of Technology, Pasadena. Andrew was my friend and skiing partner. We both enjoyed double black diamond ski runs (the most difficult) and skied together in Tahoe, Aspen, and France. A few years ago I was shocked to learn that he had died unexpectedly. We all miss him terribly.

After the early experiments that first found the Doppler peak, the next major mission to study the CMB was the Wilkinson Microwave Anisotropy Probe (WMAP) satellite. Originally named simply MAP, the letter W was added in honor of David Wilkinson, one of the founders of the field of CMB experiments. He died in 2002 shortly after the launch of the satellite. I remember him

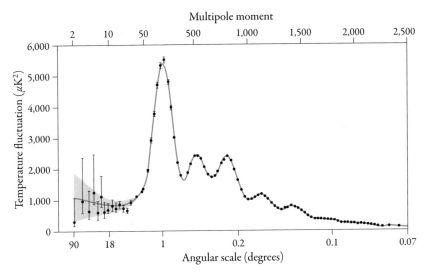

**FIGURE 3.12** The Doppler peak in the cosmic microwave background (CMB) is at an angular scale of 1 degree, exactly as predicted for a Universe with a flat geometry. A combination of information from the secondary peaks (to the right of the main one) reveals the amount of atomic matter to constitute 5% of the Universe, dark matter to make up 26% of the total, and dark energy to constitute the remaining 69%. This plot is the result of data from the European Space Agency's Planck satellite from March 2013. *Reproduced with permission from* Astronomy & Astrophysics, © ESO, *original source ESA and the Planck Collaboration, P.A.R. Ade et al. [Planck Collaboration]. "Planck 2013 Results. I. Overview of Products and Scientific Results." arXiv:1303.5062 [astro-ph.CO].*

as the thoughtful advisor of undergraduate physics majors at Princeton. In our junior year, he held a meeting to speak with us about possible futures after graduation. While I am the only one still in academics, others later went to law school, medical school, industry, banking, palm reading, and saxophone playing. It was my privilege to know Dave Wilkinson throughout my early career.

The WMAP satellite, a NASA Explorer mission, was launched in June 2001. After picking up speed from a fly-by of the moon, it headed for its final location, known as L2, or the second Lagrange point. This is one of the few places that a satellite can be placed where it will remain in an absolutely stable orbit. In the late eighteenth century, the Italian-French mathematician Joseph-Louis Lagrange discovered mathematically that there are five special points in the vicinity of any two orbiting masses.[13] At these points, a third smaller mass can stay in orbit at a fixed position relative to the other two. In the Sun/Earth system, the second Lagrange point is located directly behind Earth as viewed from the Sun, at a distance of roughly 1.5 million kilometers (almost a million miles) from Earth (Figure 3.13). Because L2 is farther away from the Sun than Earth is, objects at L2

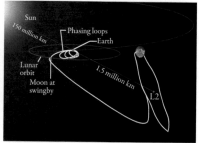

FIGURE 3.13 (Left) Data from NASA's Wilkinson Microwave Anisotropy Probe (WMAP) satellite has led to the precision era in cosmology. Yet the dark side of the Universe remains an enigma. (Right) The path taken by WMAP to get to its current location at L2, the second Lagrange point of the Sun-Earth system, where WMAP resides on a stable orbit 1 million miles from Earth. The European Space Agency's Planck satellite also resides at L2, as will the upcoming James Webb Space Telescope, the sequel to the Hubble Space Telescope. *NASA/WMAP Science Team.*

should be orbiting more slowly due to the slightly decreased gravitational pull of the Sun. Yet the added mass of Earth provides just enough extra gravity to keep the objects co-rotating around the Sun together with Earth. The WMAP satellite is currently comfortably located at L2. At all times its mirrors point away from the Sun, which is so bright that it could destroy the satellite's sensitive instruments. Future missions, including the James Webb Space Telescope and Gaia (which will provide detailed studies of a billion stars in our Galaxy) will also be located at L2.

The WMAP satellite has been a remarkably successful instrument in making detailed measurements of multiple cosmological parameters. WMAP's Charles (Chuck) Bennett, Lyman Page, and David Spergel won the prestigious Shaw Prize in 2010 "for their leadership of the WMAP experiment, which has enabled precise determinations of fundamental cosmological parameters, including the geometry, age, and composition of the Universe."[14] The WMAP team also received the 2012 Gruber Cosmology Prize.

The WMAP satellite's observations produced a full-sky map of the microwave background with images to a resolution of 0.2 angular degrees. Figure 3.14 shows the microwave sky as seen by WMAP. The image is like a fingerprint of the universe. In the color version of the figure, the hot (red) spots and cold (blue) spots are indicated by the false colors in the picture. This is our Universe as seen in the microwave. Another universe might look statistically the same, but the details seen in this picture are unique to the Universe we inhabit. The initials "SH" (Steven Hawking) can be seen in blue just to the left of center (just a joke!).

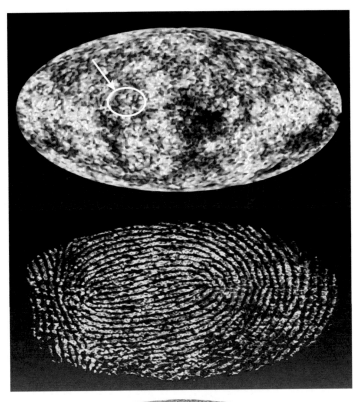

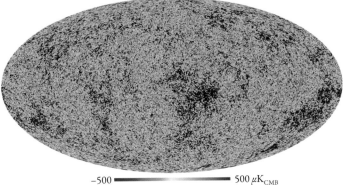

−500 ▬▬▬▬ 500 $\mu K_{CMB}$

FIGURE 3.14 (A color version of this figure is included in the insert following page 82.) The microwave sky as seen by the Wilkinson Microwave Anisotropy Probe (WMAP) (top) and Planck (bottom) satellites. Hot spots are red/orange, whereas blue regions are cold (compared to the average 2.76 K temperature). The putative initials of Stephen Hawking are circled. The WMAP and Planck images are like a fingerprint of our Universe. *(Top) NASA/WMAP Science Team; (bottom) ESA/Planck Collaboration.*

The European Space Agency's Planck satellite, launched on May 14, 2009, obtained even better data with higher resolution. After surveying the microwave sky for one and a half years, the Planck collaboration released 28 publications all on the same day, March 22, 2013. The results are beautiful. Planck's image of the microwave sky is shown in the bottom panel of Figure 3.14.[15] The data from the WMAP and Planck satellites have taught us a tremendous amount about the Universe. The most important result is the measurement of the geometry of the Universe.

Figure 3.12 illustrates the magnitude of temperature variances in the Planck data between different directions, with large angles to the left of the picture and small angles to the right. The Doppler peak is clearly at an angular scale of 1 degree, exactly as predicted for a flat Universe. The geometry is flat to within 1%.

In addition to this major peak, CMB scientists obtained a great deal of information from the smaller peaks at multiples of the primary frequency. These secondary peaks, to the right of the main one (at smaller angular scales), are caused by higher harmonics of the same signal, much as the ringing of a bell produces a variety of complementary tones. The height of the second peak determines the atomic fraction of the Universe to be roughly 5% of the total content of the Universe. This measurement confirms the two-decade-old result from studies of Big Bang nucleosynthesis that is the subject of Chapter 4. The discrepancy between the total energy density in the universe and the amount of ordinary atomic matter, seen both in the CMB and in tests of nucleosynthesis, leads to the inescapable conclusion that most of the Universe must be made up of something new and exotic.

The solid curve in Figure 3.12 shows a fit to the CMB data of a theoretical model that takes into account six important parameters describing the Universe. By varying these parameters, researchers found the values that best fit the data. As a result of this fit, astronomers determined the abundances of the dark side of the Universe—dark matter and dark energy. In Chapter 9, which focuses on dark energy, we'll see that dark matter and dark energy are completely unrelated (as far as we know). These two quantities cannot be transmuted into one another; the only thing they share is that they are both dark in the sense that they do not shine and that we do not know what they consist of. In 2009 the WMAP team found best-fit values with dark matter contributing 23.3% and dark energy making up 72.1% of the Universe. The WMAP team estimated that the possible uncertainties in these numbers were roughly 1.5%. These results became known as "precision cosmology." All of us in the astrophysics community believed that these values were accurate. Consequently, everyone was surprised when in 2013 the Planck team released results that varied significantly from the WMAP numbers. The new Planck results indicated a higher dark matter abundance and lower amount of dark energy. The current best-fit values from the Planck satellite are 26% dark matter and 69% dark energy (with uncertainties of less than 3%). These new numbers are outside of the error bars of the older WMAP values. Clearly the detailed values are still fluctuating as more data are acquired. Yet the basic consensus picture of a Universe composed of roughly 30% matter (the sum of atomic matter plus dark matter) and 70% dark energy is certain.

The CMB satellites have produced many more important results. An important number measured by Planck and WMAP is the current age of the Cosmos. They found that the Universe is 13.8 billion years old. The Planck data also confirmed several strange unexplained features previously seen by WMAP. A huge cold spot occupies approximately 5 angular degrees in the sky and is 70 micro-Kelvins (70 millionths of a Kelvin) cooler than the average CMB. Its origin is still a mystery. Another surprising result is the low magnitude of the CMB measurements on very large angular scales, a result that has caused a great deal of speculation. Yet the most likely explanation is *cosmic variance*: astronomers have only one Universe to observe, leading to statistical abnormalities on large scales. The resulting uncertainty is indicated by the thick shaded band on the lower left-hand side of Figure 3.12 at angles greater than about 1 degree. Several other anomalies in the CMB data are also intriguing. There appears to be an asymmetry in the average temperatures across the sky: the southern hemisphere is slightly warmer than the northern hemisphere. These anomalies may be giving us hints of new physics beyond the standard cosmology, or they may be just artifacts of how the measurements were made.

The results of the CMB observations have answered a key question about our Universe: What is its total mass and energy content? Due to Einstein's General Relativity, the geometry is intimately related to the mass and energy content of the Universe via the warping of spacetime. The discovery of a flat geometry determines the total mass and energy of the Universe to have a density of $10^{-29}$ grams per cubic centimeter. This amount of mass and energy is very diffuse and can be thought of as the density of outer space. For comparison, the density of water on Earth is roughly 1 gram per cubic centimeter. The contents of a single coffee cup containing the material of the average universe would weigh only about $10^{-28}$ grams, whereas the equivalent volume of coffee itself would weigh about 10 grams. If we define the observable Universe to be

FIGURE 3.15 Sir Martin Rees, the author, and Carlos Frenk.

the patch we can see (out to as far as light could have traveled in the age of the Universe), then within this patch there are $10^{55}$ grams of material.

## Outer Space and the Queen of Sweden

In 2005, the Royal Swedish Academy of Sciences awarded the prestigious Crafoord Prize in Astronomy to three astrophysicists: James (Jim) Gunn, James (Jim) Peebles, and Sir Martin Rees (Astronomer Royal of Great Britain). Swedish industrialist Holger Crafoord created this award because there is no Nobel Prize in Astronomy. The Crafoord Prize is given in both the fields of astronomy and mathematics every 3 years. (In the other years the prize rotates to geoscience, bioscience, and polyarthritis research.) The three astrophysicists won the award for groundbreaking "contributions to understanding the large scale structure of the Universe."[16]

Accompanying the awards ceremony was a small conference of 20 people. The participants were invited to give lectures on their specialties as well as to attend a banquet with the King and Queen of Sweden. I gladly accepted an

FIGURE 3.16 The potted plant: my initial attempt at taking a picture of Queen Silvia of Sweden.

invitation to join in the festivities and to speak as an expert on the subject of dark matter (Figure 3.15).

The ceremony was quite remarkable. In between presentations by King Carl Gustaf of Sweden to each of the awardees, a group of young singers performed ABBA songs on the stage. At the banquet, I was seated together with the granddaughter of the industrialist who had donated the prize. We had a great time with lots of wine. I succeeded in taking a picture of the King, but in my attempt to take a picture of the Queen at a nearby table, I accidentally took a picture of a potted plant (Figure 3.16).

After the dinner, the granddaughter of Holger Crafoord introduced me to her mother, who expressed frustration that women never won the prize.[17] She then introduced me to her brother. He asked me to take a photograph of him together with Licia Verde, another astrophysicist. At the conference, Licia was the speaker on the subject of the CMB. Unfortunately, I managed to cut Licia's head off in the picture. When the grandson saw the photos, including the one of the potted plant, he laughed and said he'd help me get better pictures. He took me up to Queen Silvia, introduced me, and took several photos of the two of us (Figure 3.17).

The origin of the current Swedish royal family is quite interesting. Over the centuries the Swedes and the Danes were perennially at war. However, when the Swedish royal line died out, the Swedes would import a Dane to become

**FIGURE 3.17** A photograph with Queen Silvia of Sweden.

King. In 1810, the crown prince to the Swedish throne died. The aging King Karl XIII wanted to elect the deceased's brother, a Danish prince, as the next heir to the throne. The story has it that a Swedish courtier was sent to Napoleon to confirm this choice. However, en route, he met one of Napoleon's marshals, Jean-Baptiste Bernadotte, a French commoner who had risen through the military ranks. The courtier became so enthusiastic about Bernadotte that he offered him the Swedish throne. Though the courtier himself was arrested on his return to Sweden, his choice gradually grew in popularity. Eventually Bernadotte was adopted by the Swedish king. He was crowned King Karl XIV Johan in 1818. Since then the House of Bernadotte has reigned as the royal house of Sweden. Today, King Carl XVI Gustaf is the descendant of this French military officer.

Queen Silvia smiled throughout the dinner and the evening. When she learned that I was a cosmologist, she told me that there was a question that had always puzzled her. She asked me how I would define space. She mentioned that she'd previously asked a mathematician and had merely become more confused. I thought about it for a second, and then decided I would define what

I thought of as "outer space." I said, "Let's imagine we're on a rocket ship. We leave the gravitational pull of Earth, and then we're out in the Solar System. Now let's keep going. Let's leave the gravitational pull of the Sun, then of the Milky Way, then of our Local Group of Galaxies. Eventually we get to an average place in the Universe, devoid of any structure. This location has the average density of the Universe, $10^{-29}$ grams per cubic centimeter. Compared with water on Earth, with a density of 1 gram per cubic centimeter, this is a very diffuse empty place. This is what I would call outer space."

Queen Silvia looked puzzled. In retrospect, I think she was referring instead to the mathematical concept of space, as in spacetime. An awkward silence followed, and I felt a mounting sense of panic. Thinking, "I must make some conversation," I stupidly asked, "Are you best friends with the Queen of Norway?" For the first time that evening, her smile vanished, and she said, "Not particularly." Later I learned that the conference coincidentally took place on the 100-year anniversary of the secession of Norway from Sweden. To quote Wikipedia: "Negotiations led to Norway's recognition by Sweden as an independent constitutional monarchy on 26 October 1905. On that date, King Oscar II renounced his claim to the Norwegian throne, effectively dissolving the United Kingdoms of Sweden and Norway." I certainly had no intention of offending the Queen![18]

## Pie Picture of the Universe

The cosmic abundances tell a consistent story. The CMB has revealed the total mass and energy content of the Universe. Together with the CMB, the element abundances from primordial nucleosynthesis (the subject of Chapter 4) restrict ordinary atomic matter to constitute only 5% of the Universe. Chapter 2 discussed the myriad pieces of evidence that dark matter is the predominant matter component. Matching all the peaks in the CMB reveals that roughly 26% of the total Universe is made of dark matter. The remaining piece, the bulk of the Universe, is made of dark energy.

The existence of dark energy was discovered in the late 1990s by two groups of astronomers studying distant supernovae, the bright explosions of dying stars. The observers found that the supernovae were significantly fainter than expected and postulated that the reason might be that they are accelerating away from us. This idea of an accelerating Universe created a paradigm shift in cosmology. We now think that more than two-thirds of the Universe consists of antigravitating dark energy. Its nature is a puzzle that leaves scientists

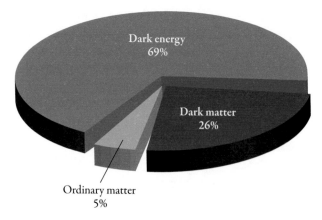

FIGURE 3.18 (A color version of this figure is included in the insert following page 82.) Pie chart of the Universe showing its three primary components.

as yet utterly perplexed. Chapter 9 is devoted to dark energy: how it was discovered, the quest to understand its nature, and its consequences for the fate of the Universe.

Our current vision of the Universe can be thought of as a pie chart containing three major components: 5% atoms, 26% dark matter, and 69% dark energy (Figure 3.18). The 5% value is known to excellent accuracy. Although the other two numbers are still fluctuating within a few percent as new instruments take data, the basic breakdown into the three components is certainly correct. Scientists are still adapting to the new reality of the dark components and hoping to fill in all the missing pieces. The vital question remains: What is the dark Universe made of?

# FOUR

# Big Bang Nucleosynthesis Proves That
# Atomic Matter Constitutes Only 5% of the Universe

The epoch of Big Bang nucleosynthesis is known as the crowning jewel of the Big Bang model.[1] When the Universe was just a few minutes old, protons and neutrons in the Universe congregated for the first time to form more complex elements that make up our world today. Beginning with early work in the 1940s, scientists predicted the abundances of helium, lithium, and other elements created at this very early time.[2] Remarkably, decades of observations found a nearly perfect match with these theoretical predictions. The success of primordial nucleosynthesis helped establish that the early Universe really did start out in a hot dense early phase. Of deep importance was the realization that the predictions work only if the amount of atomic matter in the Universe is less than a tenth of the total content of the Universe. The early researchers in this field came to the revolutionary conclusion that there must be exotic stuff in the Universe, the material we have come to know as dark matter and dark energy. These conclusions predated by several decades the more precise measurements of the cosmic microwave background radiation that nailed down the fraction of atomic matter at 5% of the total mass and energy budget of the Universe.

## A Story about Big Bang Nucleosynthesis

I was first exposed to the ideas of modern cosmology at a lecture on Big Bang nucleosynthesis during a summer internship at the Cornell Electron Storage Ring (CESR) in upstate New York. This is one of only a handful of particle accelerators, or atom smashers, on the planet.[3] At CESR (pronounced Caesar), physicists sped electrons up to nearly the speed of light and smashed them into positrons, their antiparticles. The goal was to discover new elementary particles created in the collisions. The accelerator facility housed two different experiments. The larger detector CLEO (short for Cleopatra) involved more than 40 institutions and several hundred experimentalists. I worked on the smaller CUSB detector, named for Columbia and Stony Brook, the two institutions in the group. My group had originally wanted to name the detector ASP, for the snake that bit and killed Cleopatra, the Queen of Egypt, since CLEO was the name of the competing experiment. However, someone eventually decided that this joke was in poor taste.

I was a graduate student at Columbia University and at the time planned to become a high-energy particle experimentalist. With a fellow student, I drove the 5 hours northwest up from New York City to the site of the experiment. Arriving in Ithaca after dark, we got completely lost. We found ourselves on a pig farm that was part of Cornell's agricultural college. The transition from Manhattan could not have been more extreme. We drove around randomly for about an hour in the dark farmland, hoping to find a person or even a telephone (this adventure predated cell phones). At one point, we were thrilled to find a shopping mall, only to discover that all the shops were empty. It was an eerie experience. We later figured out that it was a mall under construction. Eventually we stumbled into the town of Ithaca, where we wandered into a bar and asked hesitantly, "Is there a phone here?" People looked at us like we were crazy. The bartender said "Of course," and directed us to a pay phone. We finally reached fellow graduate students in Ithaca, who were able to direct us to our lodgings.

I had fun that summer designing and soldering pieces of electronics for the experiment and playing a lot of tennis. A few weeks after I arrived, I went to my first cosmology lecture given by a fellow graduate student. At that point I hadn't yet understood that particle astrophysics was a real science. The field was still in its infancy.

The talk blew me away. For the first time I learned that scientists were making predictions about physics that took place only a few minutes after the Big Bang, and that all the predictions proved to be right. Theorists computed that 25% of the mass of atoms in the Universe is made of helium-4, 1 part in 10,000 is deuterium, and 1 part in 10 billion is lithium—all these numbers, which differ by 10 orders of magnitude, turned out to be in perfect agreement with data. Further, the consistency of the whole picture requires the amount of atomic matter to be a small fraction of the entire mass of the Universe. There has to be exotic matter dominating the Cosmos. Big Bang nucleosynthesis, the formation of elements when the Universe was 3 minutes old, is a masterpiece of early cosmology and one of the pillars that serves as proof that the Big Bang model is correct. The lecture was a turning point for me, sparking my interest in the field of cosmology. Soon after, I moved to Fermilab, took David Schramm's cosmology class, and then transferred to graduate school at the University of Chicago to work with him as my advisor. Since then I've devoted my professional life to astroparticle physics.[4]

Let's start with a brief history of the early Universe and also describe the remarkable results I first encountered during that summer in Cornell.

## Particles in the Early Universe: A Primordial Soup

Cosmologists believe that the Universe is 13.8 billion years old. In the earliest days there was a primordial soup of subatomic particles, including quarks, electrons, neutrinos, gauge bosons, and dark matter particles—the building blocks for all of nature (Figure 4.1). The six types of quarks—up, down, charm, strange, top, and bottom—derive their whimsical name from James Joyce's *Finnegans Wake*: "Three quarks for Master Mark."[5] Most ordinary matter is made of up and down quarks and the gluons that hold them together. When additional quarks were discovered in more unusual forms of matter, they were given names like strange and charm because their existence originally came as a surprise.

Quarks combine to make the more familiar particles in nature.[6] For example, protons and neutrons are each made up of three quarks (Figure 4.2). A proton consists of two up quarks and one down quark, whereas neutrons have the reverse: two down quarks and one up. Gluons are the force carriers for the "strong force" that holds the quarks together inside the nucleus.

Associated with the six types of quarks are six types of leptons—electrons, muons, taus, and their respective neutrinos. Unlike quarks, which bind together to make more complex particles, leptons exist alone as independent entities. The most familiar of the leptons are electrons, such as those in orbit around nuclei in atoms. For every positively charged proton inside a nucleus, there is an accompanying electron with negative charge; thus atomic matter today is electrically neutral. Muons and tau particles are similar to electrons but heavier, and they rapidly decay to electrons on microsecond time scales.

The other three leptons are nearly massless particles called neutrinos, represented by the Greek letter $\nu$. These are light, electrically neutral particles. Other than gravity, the only force they feel is the weak force, which is responsible for some types of radioactivity. Neutrinos are created in nuclear reactors, in the Sun, in supernovae, and in cosmic rays. The Standard Model of particle physics contains three types of neutrinos: electron neutrinos, muon neutrinos, and tau neutrinos.

# Elementary particles

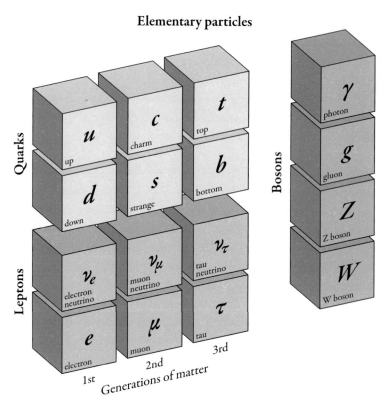

**FIGURE 4.1** Three types of elementary particles are the basic building blocks of nature: quarks, leptons, and bosons. Quarks come in six types—up, down, charm, strange, top, and bottom—and combine to make objects like protons and neutrons. Gauge bosons are the particles responsible for the fundamental forces: photons mediate the electromagnetic forces that attract electrons to the protons in atomic nuclei, gluons mediate the strong force that binds together the quarks inside protons and neutrons; and Ws and Zs mediate the weak interactions that are responsible for some types of radioactivity. These elementary particles, together with the Higgs boson, constitute the Standard Model of particle physics. *Fermilab.*

Neutrinos were first discovered in a reaction called beta decay, in which a neutron converts to a proton and in the process releases an electron. Because neutrons are slightly heavier than protons, protons are the more stable of the two objects, and this decay occurs readily in nature. Physicists studying the decay of neutrons to protons and electrons were puzzled by a surprising result. They found that the energies of the outgoing electron and proton alone didn't match up with the energy of the initial neutron. Yet energy must be conserved

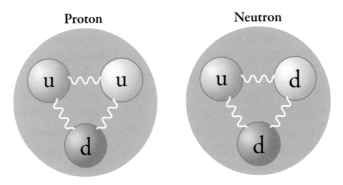

**Proton** **Neutron**

**FIGURE 4.2** The quark structure of the proton and neutron. Gluons (squiggly lines) hold the quarks together.

in every reaction. In 1930 Wolfgang Pauli had an idea that could resolve this discrepancy. In a letter to a group of physicists at a meeting in Tübingen, Germany, he began, "Dear radioactive ladies and gentlemen," and then went on to propose a "desperate remedy" to rescue energy conservation.[7] He postulated the existence of a new undetected particle as responsible for carrying off the missing energy out of the detector. This explanation proved to be correct, and these particles are now called neutrinos. The complete reaction in beta decay is $n \rightarrow p + e + \bar{\nu}_e$, where an anti–electron neutrino $\bar{\nu}_e$ is among the final products. (Antimatter is discussed in Chapter 5.)

Because neutrinos interact weakly with matter, their actual discovery took several more decades. In 1956 Clyde Cowan and Frederick Reines succeeded in detecting antineutrinos from a nuclear reactor at the Savannah River Site near Augusta, Georgia. Since their discovery, neutrinos have been studied in detail and are now regularly produced in experiments. I once had a long interesting chat with one of the discoverers, Fred Reines, at a conference dinner. The next time I ran into him, I was surprised and disappointed that he didn't recognize me. I was later sad to learn that he had developed Alzheimer's disease. Unfortunately, he didn't win the Nobel Prize for his discovery until 1995, and by that time I'm not sure he was able to appreciate the award. Scientists typically don't receive Nobel Prizes until they are quite old. What a shame that they don't receive more recognition for their work earlier! The public image of science is also affected by this fact, as young people are not often presented with reasonably young or even middle-aged role models they can relate to.

In 2012 neutrinos garnered a fair amount of interest because of claims that they might be moving faster than the speed of light! Researchers created neutrinos at the particle accelerator at CERN in Geneva and then aimed them toward a detector in Italy, the Oscillation Project with Emulsion-Tracking Apparatus (OPERA) instrument. The results of the experiment seemed to indicate superluminal speeds for the neutrinos. As an explanation, some scientists proposed that the neutrinos could travel into extra dimensions, outside the three spatial dimensions of our everyday experience. The neutrinos could take a shortcut via the extra dimensions and then return to the detectors in our ordinary world. In this scenario the neutrinos would move with ordinary speeds but travel a shorter distance through the extra dimensions, while appearing to violate the limitations of the speed of light. I had previously worked on such faster-than-light shortcut routes with Dan Chung (at that time a postdoctoral fellow at Michigan and now professor at the University of Wisconsin) in an entirely different context. We were proposing an alternative explanation to inflationary cosmology for the large-scale smoothness of the Universe that remains unexplained by the Big Bang.

If information could travel faster than the speed of light, it would lead to violation of causality. Something you do today could affect what happened yesterday. Here's a joke based on this premise. "We don't allow faster than light neutrinos in here," said the bartender. A neutrino walks into a bar.

Eventually the OPERA experimentalists realized that their results were due to a loose fiber-optic cable causing incorrect time measurements. The group retracted any claims of superluminal neutrino speeds. This is the way science works: speculative new results are presented and scrutinized, and mistakes are rectified.

Additional fundamental atomic particles in the Standard Model are the gauge bosons, the particles responsible for the forces of nature. Photons are the mediators for electromagnetic interactions, such as the attraction between protons and electrons. W and Z particles are responsible for weak interactions, such as the beta decay discussed above. Finally, gluons are the mediators for the strong interactions that hold together the neutrons and protons in nuclei as seen in Figure 4.2. The last ingredient in the Standard Model of Particle Physics is the Higgs boson responsible for giving particles their masses. The recent discovery of the Higgs is described in Chapter 6.

All these types of particles—quarks, leptons, gauge bosons, and Higgs bosons—existed abundantly in the hot early days of the Universe (Figure 4.3). They were densely packed together and roamed everywhere, moving rapidly and smashing into one another constantly. The particles were all in *thermal equilibrium*. Their numbers and their rapid movements were determined by the temperature of the Universe. Occasionally two quarks would merge

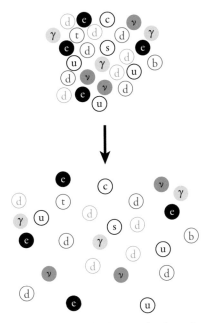

FIGURE 4.3 The early Universe was a primordial soup of quarks, leptons (such as electrons and neutrinos, symbolized by ν), and bosons (such as photons, symbolized by γ). As the Universe expanded, these particles moved farther apart and their collisions with one another became less frequent.

together, but the bombardment from other particles would break them apart again. Composites of more than one quark were not yet stable.

As time went on, the Universe expanded and cooled off. The subatomic particles spread out and slowed down, hitting one another less frequently. One ten-thousandth of a second after the Big Bang, a turning point took place, known as the quark-hadron transition. At this time quarks melded together to make bigger, more familiar particles, such as protons and neutrons. The Universe was no longer hot enough to break these particles into their constituents. Protons and neutrons became stable instead of continually being blasted apart by further collisions. Hydrogen nuclei, which combine two up quarks and one down quark into a single proton, came into existence.

At the time of the quark-hadron transition, the only atomic element in existence was hydrogen. This is the first entry in the periodic table of the elements, which contains more than 100 substances. Hydrogen has atomic number 1, indicating that is has a single proton. The formation of more complex elements, containing more than one proton, began during Big Bang nucleosynthesis.

## Origins of the Elements

Elements heavier than hydrogen began to form at the time of Big Bang nucleo-synthesis 3 minutes after the Big Bang. The temperature was then about 10 billion K, the equivalent of the temperature in nuclear reactions. The entire Universe was as dense as the material inside the nucleus of an atom today. The key event that allowed the process of nucleosynthesis to begin was the formation of deuterium atoms. Deuterium (D) is the name given to an atom that contains one neutron (n) and one proton (p) held together inside the nucleus. Because it has only one proton, D shares the same atomic number as hydrogen and is known as an isotope of hydrogen.[8] The formation of deuterium atoms was the first step in a sequence of reactions leading to the creation of elements that are more complex than hydrogen.

Deuterium formed via collisions of neutrons and protons,

$$p + n \rightarrow D + \gamma,$$

where the outgoing photon is symbolized by $\gamma$ and its presence indicates that energy is created by this reaction. Deuterium is in an energetically more stable state than the combination of an individual neutron plus an individual proton. It takes energy to split deuterium apart.

However, in the earliest times, when the Universe was not yet 3 minutes old, any deuterium that formed was short lived. When a proton and a neutron combined to make deuterium, an energetic photon from the hot primordial bath would hit the deuterium and dissociate it back into its constituent neutrons and protons:

$$p + n \rightarrow D + \gamma \rightarrow p + n.$$

The upper panel in Figure 4.4 illustrates this situation. Deuterium atoms were not yet stable. Because deuterium formation is the first step in a chain of reactions, the formation of all heavier elements was stalled. But once the temperature of the Universe dropped to 10 million K, deuterium stabilized. The ambient photons in the Universe were no longer energetic enough to blast apart the deuterium once it formed. After that, the deuterium bottleneck was broken. Nucleons (neutrons and protons) were able to stick together to make more complex elements.

Further nuclear reactions then took place almost immediately, leading to the production of new types of elements, helium and lithium. An element is defined by the number of protons it contains: hydrogen has one proton, helium has two protons, lithium has three protons, and so on. The number of neutrons can vary for each of these elements, leading to different isotopes. The

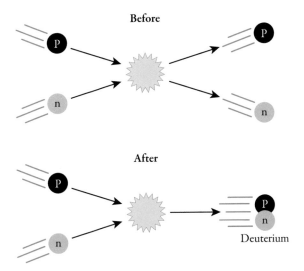

**Before**

**After**

Deuterium

**FIGURE 4.4** Before and after the first 3 minutes. The formation of elements more complex than hydrogen began with the formation of deuterium approximately 3 minutes after the Big Bang. Deuterium, which consists of one neutron and one proton, was the first building block of the other light elements. Before the Universe was 3 minutes old, deuterium wasn't stable; when it formed, it immediately dissociated again into its constituent protons and neutrons. After the Universe was 3 minutes old, deuterium became stable, triggering the beginning of Big Bang nucleosynthesis.

number after the element (for example, the "4" in "helium-4") identifies the isotope; this number, known as the atomic weight, is the total number of neutrons and protons combined. The predominant isotope of helium in the Universe is helium-4, containing two protons and two neutrons; the second most prevalent is helium-3 with two protons and one neutron.

The detailed abundances of these elements arising during primordial nucleosynthesis can be predicted with a sophisticated computer code that includes all the relevant chemical reactions. The results of the calculations predict that, at the time of nucleosynthesis, 25% (by mass) of all atomic matter in the universe ended up in the form of helium-4. Roughly 1 part in 10,000 of the atomic matter remained in the form of D and helium-3. The predicted amount of lithium-7 is 1 part in 10 billion.

For the case of helium-4, the basic calculation of the abundance is simple enough that we can illustrate it here in a few paragraphs. Of all the elements

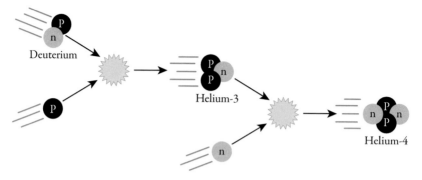

**FIGURE 4.5** A pathway for the formation of helium-4 from deuterium.

created at the time of nucleosynthesis, helium-4 is by far the most stable. It has the strongest binding energy (the binding energy is a measure of the work that would be required to tear the helium apart). Any neutrons and protons at the time of nucleosynthesis that were able to find suitable partners settled into helium-4. Using this fact, we can compute how much helium formed. Several different reaction pathways can lead to helium-4. Two deuterium atoms can merge together to make helium-4. Alternatively, a single deuterium atom can pick up a proton to make helium-3 and subsequently add a neutron to make helium-4; this is the path illustrated in Figure 4.5. Many other pathways exist as well.

To calculate the primordial helium-4 abundance, we need to know the difference in the numbers of neutrons and protons existing in the early Universe, known as the $n/p$ ratio. Here the symbol $n$ refers to the number of neutrons in the Universe and $p$ to the number of protons. At the time of the Big Bang, when all particles were in thermal equilibrium, there were equal numbers of both particles, so that $n = p$. However, neutrons are a tiny bit heavier than protons. The mass of the proton is $1.673 \times 10^{-24}$ grams, whereas the mass of the neutron is $1.675 \times 10^{-24}$ grams. The ratio of the masses is $m_n/m_p = 1.001$. Because protons are lighter, they are the more stable of the two particles. Given enough time, the neutrons would tend to convert to the lower mass protons via weak interactions, such as $n + e^+ \Leftrightarrow p + \bar{\nu}_e$; beta decay, $n \Leftrightarrow p + e^- + \bar{\nu}_e$; and various permutations, including these reactions in reverse. Here $e^-$ refers to an electron, $e^+$ to a positron, and $\bar{\nu}_e$ to the anti-electron neutrino (the antineutrino associated with the electron). Initially in the early Universe, these reactions were very rapid, and slowly but inexorably drove down the numbers of neutrons. Thus as time went on, the $n/p$ ratio slowly decreased. In the meantime, the Universe continued to expand. Eventually the particles were too far apart to continue reacting.[9] The $n/p$ ratio "froze out" to a value of $1/7$ at the time of Big Bang nucleosynthesis.

Using this ratio, we can now calculate how much helium-4 was produced at Big Bang nucleosynthesis. Because helium-4 is the most stable of all the elements formed at nucleosynthesis, as many protons and neutrons as possible settled into helium-4. Recall that a helium-4 atom has two neutrons and two protons in it, an equal number of each. Yet, with $n/p = 1/7$, the Universe contained more protons than neutrons. Thus, not all protons could find neutron companions to join up with to make helium-4. There were extra protons left over. In contrast, all available neutrons did find proton partners to make helium-4. The total number of nucleons was $n + p$, the sum of neutrons plus protons. Out of this total, only $n + n = 2n$ were able to turn into helium-4. In other words, out of the total number $p$ of protons, only those which were able to find neutron partners (that is, $n$ of them) went into helium-4. In the end, the mass fraction of helium-4 in the Universe, denoted by the symbol $Y$, was given by the ratio

$$Y = \frac{n+n}{n+p} = \frac{2n/p}{\left(\frac{n}{p}\right) + 1}.$$

Using $n/p = 1/7$ from above, we obtain

$$Y = \frac{2\left(\frac{1}{7}\right)}{\frac{1}{7} + 1} = \frac{1}{4}.$$

A simple example may be helpful. Imagine 16 nucleons (neutrons plus protons) at the time of Big Bang nucleosynthesis. To satisfy $n/p = 1/7$, on average 14 of them were protons and only 2 were neutrons. Those two neutrons joined two of the protons to make two helium-4 atoms; but 12 unpaired protons remained. So we can argue that out of the 16 original nucleons, 4 of them went into helium-4; that is, $1/4$ of the mass was converted to helium-4.

Thus we've proved it in a few paragraphs: nucleosynthesis predicts that 25% of the atomic mass in the universe was converted to helium-4 just 3 minutes after the Big Bang.

In addition, a few other elements were created during Big Bang nucleosynthesis. Calculations of these other element abundances involve a multitude of chemical reactions that are complicated enough to require a computer code known as Wagoner's code, after Robert (Bob) Wagoner, who wrote the original version. Wagoner played a seminal role in developing Big Bang nucleosynthesis. There is some residual D and helium-3: roughly 1 part in 10,000 of the atomic matter remains in the form of each of these two elements. A small amount of lithium-7 was produced as well, some via the reaction helium-3 + helium-4 $\rightarrow$ lithium-7 + $\gamma$. Another reaction mechanism produced lithium-7 via short-lived beryllium-7. Only a tiny fraction, 1 part in 10 billion of the atomic matter in the universe, ended up in lithium-7. Putting it all together, the elements produced in Big Bang nucleosynthesis were deuterium, helium-3, helium-4, and lithium-7.

Why weren't even heavier elements formed in primordial nucleosynthesis? Where did carbon, nitrogen, oxygen, or iron come from? At the time of Big Bang nucleosynthesis, the hydrogen, helium, and lithium were unable to join together to make more massive elements. A proton and a helium-4 atom might collide but were unable to form a stable atom with five nucleons in it; similarly, two colliding helium-4 nuclei could not form an atom with eight nucleons. Stable nuclei containing five or eight nucleons simply do not exist in nature. Hence no heavier nuclei could be obtained by combining two of the light ones produced in primordial nucleosynthesis. To produce anything beyond lithium, three nuclei had to merge together. For example, carbon is produced when three helium-4 atoms combine to make carbon-12. This reaction could not happen in Big Bang nucleosynthesis, because the density of the Universe at the time was not high enough for three helium atoms to interact simultaneously. This process, known as the triple-alpha process, had to wait until almost a billion years later, when stars formed. Stars are much more compact and have the higher matter densities required for three precursor atoms to merge and create heavier atoms like carbon, nitrogen, and oxygen. In order for these ingredients for life to exist, we rely on the processing of primordial elements in stars a billion years after the Big Bang. Complex elements are formed in stars and then spewed out into the interstellar medium in supernova explosions that take place once heavy stars run out of nuclear fuel. In the words of Carl Sagan, "We are made of star stuff."[10]

The predictions from the theory of Big Bang nucleosynthesis are testable. If the theory is right, the atomic matter produced in the early Universe was mostly hydrogen atoms (protons), 25% helium-4, 1 part in 10,000 helium-3 and deuterium, and 1 part in 10 billion lithium-7 atoms. These values can be compared to measurements of the cosmic abundances of these elements.

The trouble is, we don't currently live in the same chemical environment as these primordial elements. There has been a lot of processing since that time, mostly by stars. Some of the early deuterium that was created has been converted by fusion reactions in stars like the Sun into helium-4. It is hard to disentangle what we are seeing now from what was produced primordially. Our own Galaxy has had at least two generations of stars and thus a lot of chemical processing. For this reason, astronomers can't measure the primordial helium abundance by making observations of young nearby objects. Instead, the most accurate primordial measurements are made by looking at small old galaxies known as Zwicky galaxies (named after Fritz Zwicky, who first proposed the idea of dark matter—see Chapter 2). These galaxies have the lowest chemical enrichment known, and thus the values measured for these galaxies appear to be as close as we can get to those for the early Universe. By extrapolating values for the Zwicky galaxies to those appropriate to no chemical enrichment, a primordial helium-4 abundance close to 25% is obtained. This result matches the theory well. In fact, the agreement between prediction and measurement is so good that people are arguing about the second number after the decimal place in the measurement!

The primordial abundance of deuterium is even more difficult to measure. Because stars burn deuterium into helium, the amount of deuterium in the Universe today is lower than the amount that was originally produced. Between early times and now, deuterium is destroyed in many processes, never created. Thus any deuterium that is seen now places a lower limit on the primordial abundance. The best place to look for it is also outside our Galaxy. The deuterium in the intergalactic gas absorbs light from distant bright objects known as quasars, and one can see absorption lines at the appropriate wavelengths in the data. Studies of these absorption lines give a best estimate of 1 part in 10,000 (number fraction relative to hydrogen) for the deuterium abundance. Lithium-7 can be measured as well and again matches the predictions of a value of 1 part in 10 billion.[11]

Figure 4.6 shows the values of all the different abundances. The solid curves are the theoretical predictions, and the boxes indicate observed measurements. The remarkable agreement between theory and data over 10 orders of magnitude, from 25% to 1 part in 10 billion, proves that cosmologists really do understand the science of Universe almost all the way back to the Big Bang.

### Atoms Make Up Only 5% of the Universe

More can be learned from these data. As can be seen in Figure 4.6, the theoretically predicted element abundances depend on the fraction of the Universe that consists of ordinary atomic matter, also known as *baryonic matter*. The more atomic matter there is in the Universe, the higher the helium-4 abundance, and the lower the deuterium and helium-3 abundances. The rea-

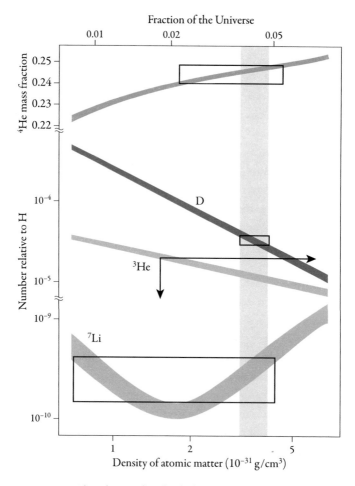

**FIGURE 4.6** Abundances for the light nuclei helium-4, deuterium (D), helium-3, and lithium-7 from Big Bang nucleosynthesis. The solid curves are theoretical predictions, whereas the boxes indicate observed measurements and the arrows indicate bounds from data. Agreement requires the atomic fraction of the Universe to be only 5% of the content of the Universe. The lower horizontal axis indicates the density of atomic matter, also known as baryonic matter. *Adapted from Burles, S., K. M. Nollett, and M. S. Turner. 1999. "Big Bang Nucleosynthesis: Linking Inner Space and Outer Space." In* Heavy Ion Physics from Bevalac to RHIC. Proceedings, Relativistic Heavy Ion Symposium, APS Centennial Meeting, *edited by R. Seto. Singapore: World Scientific, 1999.*

son for the dependence is based on Einstein's equations, which imply that the expansion rate of the Universe is faster if there is more matter. Thus, with a higher fraction of atomic matter, the interaction rates of neutrons and protons lose out to the expansion more quickly, they freeze out of equilibrium a little sooner, fewer neutrons have had time to decay into protons, and the ratio of their abundances is a little higher at the time of Big Bang nucleosynthesis. So the formula given in the previous section leads to a slightly higher prediction for the amount of helium-4. Similarly, all the other element abundances are slightly changed as well.

Already in the late 1970s there was evidence that agreement between theory and data required ordinary matter to make up less than 10% of the content of a Universe consistent with a flat geometry. Although the geometry wasn't known at the time, even then it was clear that a complete census of the contents of the Universe was likely to require the existence of something more . . . something nonatomic and exotic. Since that time, a multitude of experimental results have established that atomic matter comprises only 5% of the total content of the Universe. The first indication of the existence of nonatomic matter came from studies of Big Bang nucleosynthesis.

One of the major players in research on Big Bang nucleosynthesis was my PhD advisor David Schramm. In 1977, together with his colleagues James Gunn and Gary Steigman, Dave realized that another amazing piece of physics could be gleaned from Big Bang nucleosynthesis. They discovered that Big Bang nucleosynthesis constrains the numbers of types of atomic particles that exist in the Universe.[12] The six quarks can be collected into three families: up and down are one family, charm and strange a second family, and top and bottom make up the third family. The leptons can be categorized into the same three families: electrons, muons, and taus together with their accompanying neutrinos follow the same family structure. For decades, scientists wondered whether there could be more than three of these families. It was possible there were additional quarks that just hadn't been discovered yet. What Gunn, Schramm, and Steigman realized was that Big Bang nucleosynthesis could determine how many particle species exist. Adding additional particles to the list would cause all the curves to move to the right on the horizontal axis in Figure 4.6. The Universe would then expand too quickly at the time of Big Bang nucleosynthesis and overproduce helium-4, resulting in a prediction that is no longer in agreement with observations. The bottom line is that there cannot be more than four families in nature without violating Big Bang nucleosynthesis abundance measurements.

It is quite remarkable that astrophysics of the early Universe can serve as a probe of the fundamental constituents of nature to such accuracy. In the 1990s particle accelerators obtained more accurate measurements of the numbers of families via the lifetime of the $Z^0$ particle. This particle, one of the mediators of the weak interactions, is produced at CERN, the particle accelerator in Geneva. It decays quickly to other particles, with a lifetime that depends on the number of particle types it can decay to. Measurements of its lifetime limit the number of atomic particle species that can exist. The current picture is clear and consistent: we live in a Universe where atomic matter consists of three families of particles and no more.

Big Bang nucleosynthesis, which took place 3 minutes after the Big Bang, is a remarkable probe of the physics of the early Universe. It confirms the idea of a hot early Universe and constrains the content of the Universe today. Element abundances are consistent between theory and experiment if the Universe consists of 5% atomic matter. Further, element abundances restrict the number of atomic particle families to the three already known and understood. Any other matter, such as the dark matter that is the predominant constituent of galaxies, must be made of something new and as yet mysterious.

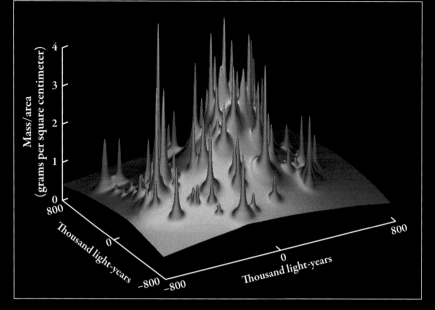

FIGURE 2.11 Computer reconstructed image of the mass distribution in galaxy cluster CL0024+1654, based on data from Hubble Space Telescope. This massive cluster gravitationally lensed the light of a more distant bright galaxy, producing multiple images of the source galaxy and allowing scientists to reconstruct the hidden mass inside the cluster. The peaks in the image are galaxies; the bulk of the mass consists of the central mountain made of dark matter in between the galaxies. *From Tyson, J. A., G. P. Kochanski, and I. P. Dell'Antonio. 1998.* Astrophysical Journal Letters *498: L107.*

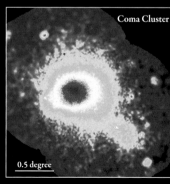

FIGURE 2.13 Galaxy cluster Coma provides evidence for dark matter. The x-rays in the image on the right are produced by hot gas, which would have evaporated from the cluster without the gravity provided by an enormous dark matter component in the cluster. (Left) Optical image. (Right) X-ray image. The two images are not on the same scale; the x-ray image focuses on the central region of the cluster. *(Left) NASA, ESA, and the Hubble Heritage Team (STScI/AURA); (right) ROSAT/MPE/S. L. Snowden.*

**FIGURE 2.14** The Bullet Cluster, a merger of two clusters each containing dozens of galaxies, gives striking confirmation of the existence of dark matter. The dark matter from lensing measurements is shown in blue; the x-ray gas composed of atomic matter is shown in red. The separation of the two components occurs because the atomic gas decelerates when it collides at the center, but the collisionless dark matter passes right on through. The existence of these two independent components is exactly as predicted in dark matter theories. The name "Bullet Cluster" refers to the striking illusion that one of the clusters looks like a bullet piercing the other. *(X-ray) NASA/CXC/CfA/ M. Markevitch et al.; (lensing map) NASA/STScI, ESO WFI, Magellan/U. Arizona/D. Clowe et al.; (optical) NASA/STScI, Magellan/U. Arizona/D. Clowe et al.*

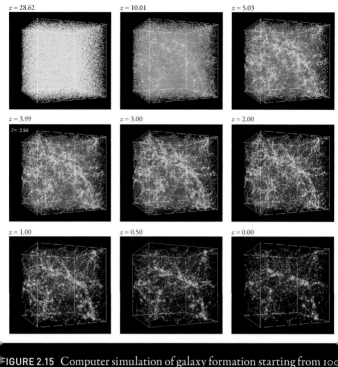

**FIGURE 2.15** Computer simulation of galaxy formation starting from 100 million years after the Big Bang ($z = 28.62$). The time sequence is labeled in terms of the redshift $z$, where higher values of $z$ correspond to earlier times in the Universe ($z = 0$ today). The bright regions in the images are actually the locations of dark matter; as the dominant matter in the Universe, it controls the formation of large-scale structure. The first small clumps of dark matter merged to form ever-larger objects, eventually creating the galaxies and other large structures we see today. Galaxies are located at the intersections of the long stringy filaments shown in the final images. Without dark matter, galaxies would never have formed and we would not exist! *Simulations were performed at the National Center for Supercomputer Applications by Andrey Kravtsov (University of Chicago) and Anatoly Klypin (New Mexico State University). Visualizations by Andrey Kravtsov.*

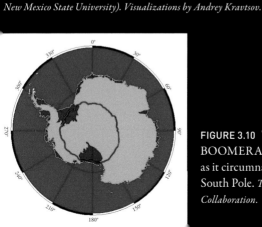

**FIGURE 3.10** The path of the BOOMERANG satellite as it circumnavigated the South Pole. *The Boomerang Collaboration.*

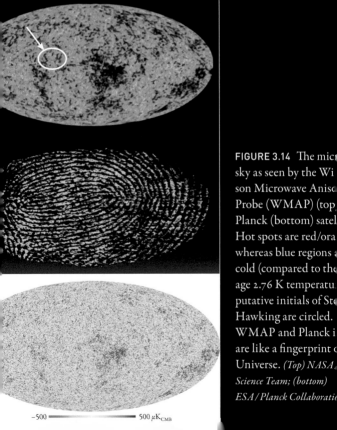

3.11  The BOOMERANG
ent about to be launched at
n Pole. In the background
uter reconstruction of the
ve images it saw. From the
he dark blue hot spots, sci-
eciphered the shape and
re of the Universe. *The*
*g Collaboration.*

FIGURE 3.14  The micr
sky as seen by the Wi
son Microwave Aniso
Probe (WMAP) (top
Planck (bottom) satel
Hot spots are red/ora
whereas blue regions a
cold (compared to the
age 2.76 K temperatu
putative initials of St
Hawking are circled.
WMAP and Planck i
are like a fingerprint c
Universe. *(Top) NASA*
*Science Team; (bottom)*
*ESA/Planck Collaboratic*

$-500$ ━━━ $500\ \mu K_{CMB}$

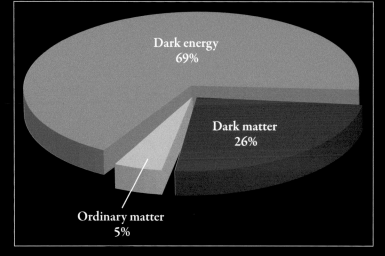

FIGURE 3.18 Pie chart of the Universe showing its three primary components.

FIGURE 6.2 Fabiola Gianotti, spokesperson for the ATLAS team of 3,000 scientists at CERN, made the first public announcement of the discovery of a new particle, most likely the Higgs boson. *CERN.*

FIGURE 6.4 (Top) The CMS detector at CERN during construction. (Bottom) Peter Higgs, who won the 2013 Nobel Prize in Physics for predicting the existence of the Higgs boson, in front of the CMS detector. *CMS and CERN. Copyright © 2008 CERN.*

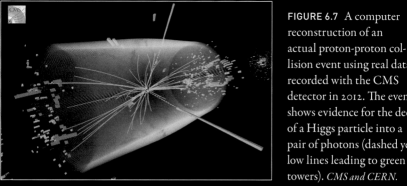

**FIGURE 6.7** A computer reconstruction of an actual proton-proton collision event using real data recorded with the CMS detector in 2012. The event shows evidence for the decay of a Higgs particle into a pair of photons (dashed yellow lines leading to green towers). *CMS and CERN.*

"I'm a Spaniard caught between two Italian women."

**FIGURE 8.4** "I'm a Spaniard caught between two Italian women." Juan Collar of CoGeNT (center), Rita Bernabei from DAMA (left), and Elena Aprile from XENON (right) are leaders of three of the principal dark matter experiments. *(Left) Rita Bernabei, ONFS; (middle) Juan Collar and the University of Chicago; (right) Richard Perry /New York Times.*

**FIGURE 9.1** A supernova remnant. *(X-ray) NASA/CXC/ SSC/J. Keohane et al.; (infrared) Caltech/ SSC/J. Rho and T. Jarrett.*

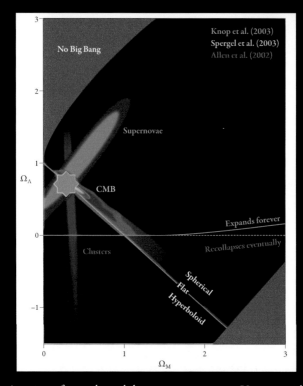

**FIGURE 9.3** A variety of astrophysical data sets converge on a Universe with 31% matter and 69% dark energy. The horizontal axis is the fraction of the Universe consisting of matter (both atomic and dark matter), and the vertical axis is the fraction consisting of dark energy. The green regions are consistent with supernova data; the red regions with cluster data; and the brown regions with cosmic microwave background (CMB) data. The best fit that matches all the data (indicated by a star) is 31% matter and 69% dark energy. Although this figure doesn't make the distinction, the matter content is further divided into 5% atomic matter and 26% dark matter.

**FIGURE 9.7** "Dark matter is attractive, while dark energy is repulsive!"

# FIVE

# What Is Dark Matter?

The nature of 85% of the content of the Universe is a mystery. The name "dark matter" attests to the fact that it doesn't give off light. Yet it provides the gravitational pull to hold together the galaxies and clusters we observe in the Universe today. We see its gravitational effects everywhere—it bends light; it makes gas whizz around the centers of galaxies. Without it galaxies could never have formed. But what is dark matter made of? It must be made of something other than bright stars, but what?

Some of the earliest dark matter candidates scientists considered were objects that are known to exist: rocks, dust, antimatter, neutrinos, faint stars, or black holes. This chapter begins with a discussion of the various possibilities and then turns to the modern view.

## Rocks or Dust

In the mid-1980s, Dennis Hegyi and Keith Olive studied the possibility that dark matter consists of gas, snowballs, rocks, or dust. First they looked at objects made purely of hydrogen, the most abundant element in the Universe, but each of the candidates turned out to have a problem. Small frozen lumps of hydrogen (snowballs) evaporate. Diffuse hot gas would produce x-rays that are not observed. Cool, neutral hydrogen should absorb background light from bright distant objects known as quasars, but not much of this absorption takes place.

Next these physicists investigated rocks and dust, which contain elements more complex than hydrogen. If the predominant matter in the Universe was composed of large numbers of these objects, then all stars should have been contaminated with significant abundances of metals. Yet the earliest stars are found to consist almost entirely of hydrogen. The simplest objects of our daily experience—rocks, gas, and dust—do not exist in sufficient quantities to solve the dark matter problem.

I will always feel a bond with one of the authors of this work, Keith Olive, because we shared the same early career path—though he was a year ahead of me. We both got our PhDs with Dave Schramm, the giant of cosmology, and then took the same postdoctoral fellowship at Harvard. After that he moved on to CERN in Geneva, while I moved to Santa Barbara.[1]

Since the time of the work of Hegyi and Olive, it has become clear that atomic matter of any kind simply cannot add up to explain all the dark matter. In the last chapter, we saw that primordial nucleosynthesis together with cosmic microwave background (CMB) observations restrict the atomic abundance to at most 5% of the total content of the Universe, whereas dark matter makes up roughly 26%. Thus something more exotic is needed to solve the dark matter problem.

## Matter and Antimatter

For every type of particle in the Universe, there is a corresponding antiparticle. This antimatter is known to exist throughout the Universe. One might be tempted to ask: What is the global abundance of antimatter? Can it solve the dark matter problem?

The nomenclature is straightforward. For most antiparticles, scientists simply add the prefix "anti" to the name of the particle and put a bar over the top of its symbol. The antipartner for the proton is the antiproton ($\bar{p}$), for the quark is the antiquark ($\bar{q}$), and for the neutrino is the antineutrino ($\bar{\nu}$). The exception to the naming rule is the positron ($e^+$), which is the antipartner for the electron. Antimatter consists of antiquark constituents in the same way that matter consists of quarks. Where a proton consists of two up quarks and one down quark, an antiproton is made of two anti-up quarks and one anti-down quark.

Particles and antiparticles are like cousins (or, as some say, like evil twins). They have the same mass and spin as their partners, but they are opposite in every other way. They have opposite electric charges, opposite color charges (which determine their behavior under the strong forces responsible for holding nuclei together in atoms), and the opposite values of a variety of other properties known as quantum numbers. Whereas an electron ($e^-$) has negative electric charge, a positron ($e^+$) has positive electric charge. Whereas a proton has positive electric charge, an antiproton has negative electric charge. Photons are their own antiparticles and have zero electric charge and no color charge.

Because particles and antiparticles have the same mass, they weigh the same amount and respond identically to the force of gravity.

There is no question that antimatter exists. Although it is often the stuff of science fiction, it is also real and is well studied on a daily basis in science experiments. High-speed cosmic ray particles from the Galaxy are one source of antimatter here on Earth. When cosmic rays enter Earth's atmosphere, they collide with nuclei (mainly oxygen and nitrogen) and create a cascade of particles and antiparticles. Protons impinging on the atmosphere produce pions that then decay to muons, antimuons, neutrinos, and antineutrinos; the muons and antimuons further decay into electrons and positrons. Roughly 10,000 particles per square meter arrive at the surface of the Earth every second, with a ratio of one antimatter particle for every 10,000 ordinary particles. These are easily detectable. In an undergraduate laboratory class at Princeton University, my lab partner and I were able to perform an experiment to measure the lifetime of incoming muons from the Galaxy. Astrophysicists track cosmic rays with a variety of ground-based detectors as well as with balloons or spacecraft. These make interesting probes of dark matter, as we'll see in Chapters 7 and 8.

Antimatter can also be produced here on Earth. Particle accelerators regularly make antiparticles for high-energy physics experiments. For example, at the Tevatron accelerator at Fermilab, physicists created antiprotons, accelerated them to high speeds, and smashed them into an opposing beam of protons. The Tevatron was the premier high-energy physics laboratory until it was shut down in 2009 to make way for the new world leader at CERN.

When particles meet their antipartners, they annihilate; that is, they lose their original identity and turn into something different. The word "annihilation" somehow has the connotation that the matter-antimatter pair simply disappears into nothingness, but this is not correct. Instead, when the original particles are gone, new ones appear.

In every interaction, energy, momentum, electric charge, and many other quantum numbers must be conserved. Annihilating particles and antiparticles are so well matched up that all their properties cancel out and something radically new can be produced. Because the electric charge and the other quantum numbers of the annihilating particles are automatically equal and opposite, they add to zero. The outgoing particles emerging from the collision must then also have electric charge (and other properties) adding to zero. In addition, energy must be conserved: the energies of the outgoing particles have to equal the energy of the incoming particle-antiparticle pair. The result of the annihilation could consist of either a new particle-antiparticle pair or of two photons (which are their own antiparticles). For example, Figure 5.1 shows the

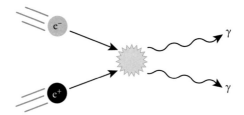

FIGURE 5.1 Annihilation of positron-electron ($e^+$-$e^-$) pair into photons ($\gamma$).

annihilation of an $e^+$-$e^-$ pair into a pair of photons. Alternatively, the annihilation products could be a quark-antiquark pair, a pair of W particles (discussed in the previous chapter), or a neutrino-antineutrino pair. At the Tevatron, collisions of protons and antiprotons led to the discovery of the top ($t$) quark in 1995 via the reaction:

$$p + \bar{p} \rightarrow t + \bar{t}$$

Such annihilation mechanisms may someday be used as a power source. It already powers the engines of fictional starships, such as the Enterprise in *Star Trek*. In Dan Brown's novel *Angels and Demons,* a canister of antimatter stolen from CERN and hidden in Vatican City threatens to detonate within 24 hours. The fallacy of the science fiction is that it is impossible to store antimatter in this way, because it would immediately annihilate with its container.

An important unresolved puzzle in particle physics is *matter-antimatter asymmetry*. The Universe clearly contains far more particles than antiparticles, or we would see the matter we consist of annihilate away. To date, NASA has not lost a space probe due to annihilation with antimatter. More detailed studies have found that there are a billion particles for every antiparticle in the Universe. The origin of this matter-antimatter asymmetry is perplexing and is the topic of ongoing research. Many explanations have been attempted, but as yet no resolution to this problem has been found.

When I was in graduate school, the prevailing wisdom was that baryogenesis (the creation of the observed overabundance of matter) was basically understood. Particle theorists believed that the matter-antimatter asymmetry was likely generated at a "GUT transition" in the early Universe, where GUT stands for Grand Unified Theory. This is the transition from the earliest times, when all the forces of nature were unified into one, to the split-up into separate electroweak and strong forces. Decays of superheavy particles during the GUT era could in principle have produced an excess of matter over antimatter.[2]

Unfortunately, these models of baryogenesis failed. The simplest GUT models they were based on turned out to be incorrect. These theories had predicted that protons should decay on very long but measurable time scales, but experiments proved these ideas false. Ironically, today there is no single predominant theory of baryogenesis. A plethora of good ideas is out there, but no conclusive evidence shows that any one of them is correct.

Antimatter is not the same thing as dark matter. The antiparticles of ordinary matter are still part of the atomic content of the Universe. In the pie picture of the Universe, we saw that there were three major components: atomic matter, dark matter, and dark energy. If we add up all the quarks, electrons, protons, and so forth together with their antimatter (antiquarks, positrons, antiprotons), their sum still makes up only 5% of the total content of the Universe—the atomic portion of the pie. The remaining two pieces of the pie are the dark matter (26%) and dark energy (69%).

Although we have only discussed the antipartners to ordinary matter, dark matter particles also have their own antimatter.[3] If we add up the dark matter particles plus their corresponding antimatter, they sum to 26% of the total content of the Universe. The dark matter portion of the pie is completely separate from the atomic matter portion (which contains antiquarks, antiprotons, and so forth). Dark matter is not the antimatter for ordinary matter but instead must be something new.

## Neutrinos

Originally discovered in 1956, neutrinos are light, electrically neutral particles. Their name is the Italian for "little neutral one." As seen in Chapter 4, other than gravity, the only force they feel is the weak force, which is responsible for some types of radioactivity.

For decades, neutrinos were considered to be among the best candidates for the dark matter of the Universe. They don't count in the 5% atomic piece of the inventory of the Universe. Because they only interact via the weak force, neutrino reactions had stopped by the time of nucleosynthesis and did not affect the abundances of primordial elements. Thus their numbers are not constrained by the restriction on atomic matter from Big Bang nucleosynthesis. Of all the nonatomic candidates for dark matter, neutrinos were the least exotic. After all, they are known to exist.

When cosmologists computed the relic abundance of these particles left over from the Big Bang, they found encouraging results. The number of neutrinos should be substantial, roughly 400 per cubic centimeter, or thousands in a coffee cup. With this large quantity, neutrinos could be light and still add up to a major contribution to the masses of galaxies. The right value of the neutrino mass to solve the dark matter problem would have been about 100 electron volts (eV), roughly 10 million times lighter than a proton. From now on we'll follow the convention of particle physicists and use electron volts, the energy equivalent of the mass unit, for particle masses. In these new units, the proton mass is 1 GeV, or 1 billion eV. If neutrinos weighed 100 eV, the neutrino contribution to the mass density of the Universe would have been just perfect for dark matter.

Speculation about neutrino dark matter abounded. Yet in the 1980s computer simulations of galaxy formation with neutrinos began to find problems with this candidate for dark matter. Because neutrinos are so light, they traverse long distances at nearly the speed of light. If they dominated the mass of the Universe, they would drag other matter along with them. As a consequence, they would prevent atomic matter from aggregating into galaxies. Structures could not have formed quickly enough to agree with observations of early galaxies. Neutrino dark matter would have interfered with the process of galaxy formation. This is known as the hot dark matter problem. Any relativistically moving (or "hot") dark matter particles such as neutrinos could not have succeeded in forming the galaxies we see today. In the very first paper I wrote, I discussed the properties of neutrinos required to allow the biggest superclusters in the Universe to form.

At the same time, experimentalists were trying to measure the mass of the neutrino. For decades Raymond (Ray) Davis saw a deficit of neutrinos from the Sun in his data—a third as many as expected from solar theories. This discrepancy came to be known as the solar neutrino problem. The proposed resolution was that some of them had converted to a different type, or flavor, of neutrino not detectable in his experiment. Such an oscillation would only occur if neutrinos were massive. My thesis advisor Dave Schramm always had good intuition about these things. He was sure that the Davis experiments meant that neutrinos had to have mass. In contrast, I didn't believe it. I tend to be a skeptic. I thought it was just uncertainties in the solar models. The numbers of neutrinos emerging from the Sun depend on its central temperature to the fourteenth power, and no one had accurate enough models to produce trustworthy predictions. Or so I thought.

In 1998, researchers at the Super-Kamiokande experiment in Japan proved Davis right. The Japanese researchers announced a definitive discovery of neutrino mass. Their detector consisted of a 50,000 ton tank of ultra-pure water located in an old zinc mine under the Japanese Alps. The experimentalists compared the numbers of muon neutrinos coming from different directions. They found fewer muon neutrinos traveling through Earth en route to the detector compared to those coming directly from the atmosphere overhead. Just as in the solar neutrino problem, the only sensible explanation was that the neutrinos traveling through Earth converted (or "oscillated") into a different type of neutrino (probably the tau neutrino). The discovery that neutrinos oscillate from one type to another implies that they have a nonzero mass.

The discovery of neutrino mass has profound consequences for particle physics. All prior elementary particle data were consistent with the Standard Model of particle physics. However, this model predicted that neutrinos should be massless. Thus the Super-Kamiokande measurement is the first evidence of unexpected new physics beyond the Standard Model. Understanding the origin of neutrino mass and explaining why it is so small will require exploration of novel theoretical ideas. In 2002 Davis and Masatoshi Koshiba received the Nobel Prize for the discovery of neutrino mass.

These experiments actually determined the mass differences between neutrinos, rather than the masses themselves. The value of the mass difference was found to be very small, roughly in the range 0.01–0.1 eV, or 10–100 billion times less than the mass of a proton. Many other experiments further constrain neutrino masses. A combination of cosmological data, including the CMB radiation and galaxy surveys, require the summed masses of the three neutrino types to be less than 1 eV. As mentioned above, the right value of the neutrino mass to solve the dark matter problem would have been about 100 electron volts. Consequently, neutrinos are simply too light to provide the dark matter. Neutrinos only add up to 0.5% of the total content of the Universe. Although the discovery of neutrino mass is extremely important to particle physics, unfortunately its value is so small that neutrinos play a negligible role in cosmology. None of the three species of neutrinos can solve the dark matter problem in the Universe.

### Sterile Neutrinos

The three weakly interacting neutrino species of the Standard Model—the electron neutrino, muon neutrino, and tau neutrino—have not survived as dark matter candidates. There is as yet one remaining window for possible neutrino dark matter: *sterile neutrinos*. This would require a new neutrino species, in addition to the known three. The name "sterile neutrinos" originates from the fact that these particles would not interact via any of the fundamental interactions of the Standard Model. The only way their existence would be felt by standard particles would be by their gravitational effects. They could live side-by-side with ordinary particles and provide the dark matter in galaxies. Future tests of this idea will have to determine how this possibility plays out—but it is not my favorite candidate.[4]

## MACHOs

For decades many astronomers favored stellar objects as the most likely explanations for dark matter. Bright stars clearly did not exist in sufficient abundance to solve the problem. Yet the possibility of a plethora of fainter objects, not yet visible in existing telescopes, held out promise as a solution. These fall into the category of Massive Compact Halo Objects (MACHOs). The name "MACHO" clearly differentiates these from the elementary particle candidates for dark matter, such as neutrinos. The MACHO class includes faint stars, planetary objects, and stellar remnants—white dwarfs, neutron stars, and black holes.

The argument for faint stars as dark matter was a sound one. As telescopes improved, astronomers could observe dimmer and dimmer objects. The fainter the stars they were able to see, the more they found. The possibility remained that there was an enormous number of really dim objects, just below the limits of detection. For decades many scientists believed that faint stars would solve the dark matter problem.

The Hubble Space Telescope (HST), carried into orbit by a space shuttle in 1990, played a major role in addressing this question. Named after Edwin Hubble in honor of his discovery of the expanding Universe, HST allowed astronomers to look farther back into the reaches of the distant Universe than ever before. HST could search for extremely faint objects. In the mid-1990s two groups used HST data to estimate the numbers of faint stars in the Milky Way.[5] My graduate student David Graff and I carefully added up the contributions of observed stars and concluded that faint stars could constitute at most 1% of the mass in the Galaxy.

Graff and I went on to investigate other MACHO candidates. Substellar objects known as brown dwarfs were intriguing possibilities. These are made of the same material as stars, but they are less massive and consequently don't produce much light. Brown dwarfs weigh less than a tenth of the mass of the Sun. They are so faint that even HST would not be able to detect them.

Graff argued that the best way to tackle the abundance of brown dwarfs is to extrapolate from the numbers of heavier stars that do shine. Faint stars and brown dwarfs are not fundamentally different. The only distinction is that the more massive ones ignite fusion and light up as stars, whereas the lighter brown dwarfs do not. Observers had found that as the mass goes down, the numbers of objects go up. Theorists argued that a simple extrapolation of the numbers versus mass was sensible even down into the brown dwarf regime.[6] Graff used this logic to extrapolate from existing data of faint stars down to the substellar regime.[7] We found that brown dwarfs constitute at most a few percent of the mass of the Galactic halo.[8] We determined that faint stars and substellar objects together add up to less than 3% of the mass of the Milky Way and do not solve the dark matter problem.

The remaining MACHO candidates for dark matter were stellar remnants. When stars die, they become one of three types of objects: white dwarfs, neutron stars, or black holes. To describe stellar remnants, a brief history of the lives of stars is in order.

### The Life of the Sun

Our Sun is currently undergoing fusion: four hydrogen atoms fuse together to make helium, releasing heat and light in the process. This hydrogen-burning phase in the lifetime of a star is known as the main sequence phase. The Sun is now 4.5 billion years old and is halfway through its life cycle. As time goes on, more and more of the hydrogen will burn to helium. Inert helium, the product of the fusion process, is accumulating at the center of the star. Eventually, in about another 5 billion years, the hydrogen will run out. The lifetime of the Sun from its creation to its demise will be close to 10 billion years.

Once the hydrogen fuel in the Sun's core is exhausted, it will lose the pressure support supplied by the fusion process. The core will contract because of gravity. This collapse will heat up the surrounding layers and ignite the hydrogen in a shell around the core. Next the helium in the center will start to burn, with three helium atoms fusing to make carbon. The Sun will puff up to become a large *red giant* star encompassing all the material in the inner Solar System, probably including Earth. Our planet will become fiery hot and may even be swallowed inside the red giant. Earth will be a deadly home for any remaining human inhabitants.

### Stellar Remnants

After the helium fuel is exhausted, the red giant phase will end. Once the fusion stops, there is nothing to prevent the star from gravitational collapse. The Sun will condense into a small dense object known as a white dwarf. The pressure support for a white dwarf is based on the Pauli exclusion principle from quantum mechanics, which says that no two electrons can inhabit a given state. When the mass of a star is condensed into a white dwarf, the electrons are so tightly packed that they fill up all the available low-energy states and push back against being squeezed any further. The resulting pressure, known as *electron degeneracy pressure,* stabilizes the white dwarf and prevents it from collapsing further.

When the Indian physicist Subramanyan Chandrasekhar first proposed the existence of white dwarfs in the 1930s as a natural consequence of the Pauli principle, his work was attacked by the influential British scientist Sir Arthur Eddington. This controversy delayed white dwarfs from being taken seriously for quite some time. Despite the objections of Eddington, Chandrasekhar was eventually proven right and in 1983 was awarded the Nobel Prize (with William A. Fowler) "for his theoretical studies of the physical processes of importance to the structure and evolution of the stars."[9] The final fate of our Sun will be to collapse to a white dwarf star. The white dwarf will weigh roughly half as much as the Sun, and yet its radius will be comparable to that of Earth. Thus the white dwarf will be about a million times as dense as Earth is. The core of the white dwarf will consist of a crystalline lattice of carbon and oxygen, a bit like a diamond.

White dwarfs are among the densest forms of matter in the universe. Only neutron stars and black holes have this much matter packed into an even smaller region. Stars that are more massive than the Sun during their hydrogen burning stages have final fates as neutron stars or black holes.

A star that is three times as heavy as the Sun will eventually collapse to a neutron star. Its mass is sufficient to make the electrons and protons merge together to form neutrons, and it is then neutron degeneracy pressure that holds up the star against further collapse. Similar to the electrons in a white dwarf, neutrons have to obey the Pauli exclusion principle: no more than one neutron can inhabit a given state. The neutrons are so tightly packed in the interior of a neutron star that they resist further compression and stabilize the star.

When stars heavier than 20 solar masses run out of nuclear fuel in their cores, they succumb to the force of their own weight and become black holes about as massive as the Sun. These objects are so compact that they fit inside a radius of 3 kilometers (5 miles). Black holes are fascinating objects and are the focus of the next section of this chapter.

In all, there are three possible outcomes for the deaths of hydrogen burning stars. We've seen that white dwarfs are the final end-products of solar mass stars such as our Sun; neutron stars are the end-products of stars of at least 3 solar masses; and black holes are the end-products of stars more massive than 20 Suns. Despite their differences, white dwarfs and neutron stars are both expected to have roughly the mass of the Sun—but with this mass compressed into a much smaller region. The typical radius of a white dwarf is 6,400 kilometers (4,000 miles), roughly the radius of Earth. The radius of a neutron star is only 10 kilometers (6 miles).

## Stellar Remnants as MACHO Dark Matter

Galactic halos could contain large numbers of white dwarfs, neutron stars, or stellar black holes. All three types of stellar remnants belong in the category of MACHOs from the point of view of the dark matter problem.

In a series of papers, David Graff, Brian Fields, and I looked for evidence of a large white dwarf population inside the Milky Way. We tested this hypothesis by looking for evidence of their progenitor stars; prior to becoming white dwarfs, a large number of precursor stars would have existed, powered by hydrogen fusion. We soon discovered that none of the expected signatures of these progenitors existed in the data. They would have produced a huge amount of infrared radiation, but it is not observed. They would have produced substantial abundances of carbon, nitrogen, and helium—again in excess of what is observed. They would have required an enormous mass budget; virtually all atomic material in the Universe would at some point have cycled through the progenitor stars. Neutron stars and stellar black holes had similar problems. In short, we did not find evidence of a stellar remnant population comprising most of the mass of our Galaxy, and we concluded that the bulk of dark matter in galaxies must consist of something else. However, we did find that white dwarfs could solve part of the dark matter problem of the Milky Way. As much as 15% of the Galactic mass could be made of a population of white dwarfs. This amount would not violate any observational tests.

At the time of our work, *microlensing* experiments named MACHO, OGLE, and EROS disagreed with us.[10] Based on their early data, these groups initially thought they had discovered sufficient numbers of MACHOs to solve the dark matter problem. They pointed their telescopes at a small nearby galaxy known as the Large Magellanic Cloud, visible only from the Southern Hemisphere. This galaxy is named after Ferdinand Magellan, who sighted it when he crossed south of the equator on a voyage in 1519. The microlensing collaborations imaged millions of galaxies in the Large Magellanic Cloud every night to look for MACHOs.

I remember sitting in the office of fellow graduate student Arlin Crotts at the University of Chicago when he first came up with the idea of microlensing.[11] At the time the idea seemed optimistic if not farfetched, because it required nightly observations of millions of galaxies. Yet only a few years later, three different groups successfully designed experiments to do these studies.

FIGURE 5.2 MACHOs are dead. Desperately seeking WIMPs. *Simon Strandgaard.*

When a MACHO in our Galaxy passes in front of a background star in the Large Magellenic Cloud, the background star looks brighter. Thus the signature of a MACHO in the microlensing experiments would be a temporary amplification of a star in the Large Magellenic Cloud. The reason for the brightening is gravitational lensing: the MACHO bends the light from the background star and focuses it onto the telescope.[12] Because it takes a MACHO anywhere from a few days to a few months to traverse the line of sight to the Large Magellenic Cloud, the timescale for a microlensing event varies in this range.

As more data came in, the experimentalists revised their estimates. Though at first they thought they had discovered dark matter, their final results were consistent with our calculations. Of course there is nothing better than an actual observation to test a theory. The results of the microlensing experiments refuted the hypothesis that MACHOs could be 100% of the dark matter in a range of masses from that of the Sun all the way down to the mass of the Moon. The only remaining possibility, as mentioned above, is that our Galactic halo could consist of up to 15% white dwarfs.

Regarding the dark matter problem, my conclusion is that "MACHOs are dead. I am desperately seeking WIMPs" (Figure 5.2).[13] On February 29, 2000, James Glanz wrote an article (which included a photograph of me at the UCLA Dark Matter Meeting that year) for the *New York Times* titled "In the Dark Matter Wars, WIMPs Beat MACHOs."[14]

## Black Holes

Black holes are the most compact and extreme objects in the Universe. Spacetime around them is heinously curved and gives rise to remarkable effects. The mass of a black hole is so compressed and its gravity so strong that any object venturing too close is trapped. Nothing, not even light, can escape. Black holes bend spacetime into a funnel shape that inexorably draws objects inward. Once inside there is no further communication with the outside world ever again.

### Black Hole Event Horizon: The Boundary of No Escape

Anything that falls inside the event horizon of a black hole is trapped forever. This is the boundary of no escape. For a black hole as massive as the Sun, the radius of the event horizon is 3 kilometers—tiny compared to the radius of our Sun, which is 10,000 kilometers. The mass of the black hole has been compressed into an extremely small region and exerts an extremely powerful gravitational pull on any object venturing too close.

A simple analogy for the gravity of a black hole is given by a rocket launch. If the rocket takes off too slowly, it will reach a maximum altitude, turn around, and be pulled back down to Earth. To escape Earth's gravity, the rocket must be propelled to at least 11 kilometers per second (25,000 miles per hour), a speed known as the escape velocity. Then the rocket can successfully leave the planet and fly out into the Solar System. If we replace the surface of Earth with the event horizon of a black hole, nothing can get away from its gravitational pull. The escape velocity would have to exceed the speed of light—but that is impossible. Even a projectile moving outward from the black hole at the speed of light turns around and is sucked back inside.

Once something falls inside the event horizon, it is inexorably pulled closer and closer toward the core of the black hole. At the absolute center is a *singularity,* or point of infinite density. Mathematically, this singularity has much in common with the moment of the Big Bang. Both are characterized by an enormous density and a breakdown of the laws of physics. Physicists hope to glean secrets about fundamental physics by studying the aberrations of these singularities. Over the past decade, string theory has made substantial progress. The centers of black holes take theorists beyond Einstein's General Relativity and toward a deeper understanding of the Universe.

The fate of a man falling into a black hole is a fun (though fatal) story. He is gravitationally attracted toward the center of the black hole just as we are attracted toward the center of Earth. Yet the gravity of the black hole is so much more enormous. As a consequence, the tiny differences of the forces on the various body parts manifest by causing his body to stretch. His feet, which are closer to the black hole, are pulled in more forcefully than his head. His right and left sides are pulled closer to one another as he approaches the black hole. The man is stretched into a thin piece of spaghetti. His atomic structure is pulled apart by these forces. Even if he were to survive the fall into the black hole, in the end his atoms would be disrupted into a primordial soup as he approached the ultimate singularity.

From an outside observer's point of view, weird phenomena take place. Let's imagine a married couple: the husband falls in while the wife watches. Imagine that this particular husband is Superman and his body can survive the strong gravity of the black hole. As the man falls into the black hole, the wavelength of his speech gets longer and longer. His voice gets lower and lower and then disappears altogether once he crosses the event horizon. The wife can berate him all she wants and won't hear anything back in return. He can hear her, but he can't communicate any response back out past the horizon. No matter how loudly he shouts, she can't hear him. As far as the woman can tell, her husband has simply disappeared into the black hole.

## Supermassive Black Holes

Black holes exist throughout the Universe. Although it is impossible to see them directly because no light can ever escape a black hole, the indirect evidence for them is incontrovertible.

At the center of every galaxy is a supermassive black hole with a mass ranging from 1 million to 10 billion times that of the Sun. Andrea Ghez and Reinhard Genzel proved that the Milky Way has a 4-million-solar-mass black hole at its center (affiliated with the strong radio source Sagittarius A*). They found stars and dust swirling around the Galactic Center at breakneck speeds of 1,000 kilometers per second (2 million miles per hour). These speeds can only be explained if the material is in an accretion disk around a black hole—a rotating flattened region, like a pancake, containing hot ionized gas that is en route to being swallowed. In 2012 these two scientists were presented the Crafoord Prize in Astronomy by the Royal Swedish Academy of Sciences "for their observations of the stars orbiting the galactic centre, indicating the presence of a supermassive black hole."[15] Astronomers can see further evidence for astrophysical black holes by studying the hot gas swirling around them. As the material in the accretion disk falls inward, it gets compressed and heated. The gas gives off radiation that escapes in enormously energetic jets that can be observed in a variety of wavelengths of light.[16]

## Black Hole Formation

Radically different formation mechanisms produce three different types of black holes. The first category consists of supermassive black holes found at the centers of all galaxies, including the Milky Way. The origin of these giants is an active area of research. One possibility is that small seed black holes (for instance, containing 100 solar masses) were created early on and then grew larger by merging together or by accreting mass from their surroundings. The trouble with this scenario is that the growth rate appears to be too slow. Astronomers have discovered billion-solar-mass black holes that already existed 9 billion years ago. Starting from small seed objects, there is not enough time to produce these ancient supermassive black holes by mergers or accretion.

My collaborators and I proposed a new type of star called a "dark star" as the explanation. If we are right, the first stars to form in the Universe were powered by dark matter annihilation rather than by fusion. Although these dark stars were made primarily of ordinary hydrogen and helium, the dark matter inside them could serve as their power source. Because the stellar surface remained puffy and cool, atomic matter was able to continue falling onto the star. Dark stars could grow to enormous sizes, as big as 10 million solar masses. When these stars died, they collapsed to black holes that could serve as precursors for supermassive black holes. Million-solar-mass seed black holes are a much better starting point for forming the giant black holes observed in the Universe.

The second category of black holes consists of those resulting from collapse of massive stars. The numbers of these are uncertain. Many could exist in galaxies. Yet, because they formed from the collapse of ordinary atoms, their abundance is restricted by the bounds on the amount of atomic matter in the Universe. They can contribute at most a few percent of the mass in the Universe, which is not enough to account for all dark matter.

The third possibility for black hole formation is far more speculative. Very early in the history of the Universe, before it was even 1 minute old, primordial black holes could have formed.[17] Small superdense regions of the Universe could have pinched off from the general expansion and collapsed into black holes. The gravitational force in these small regions would overcome the pressure, and the whole region would have simply collapsed. Typically, these black holes would be very small, a fraction of the mass of the Sun. Yet there could be so many of them that they add up to explain the dark matter in galaxies. Earth-mass black holes, for example, could be abundant in galaxies and yet could easily escape detection. Of all the types of black holes, only these primordial black holes remain possible dark matter candidates.

## Black Hole Collisions and Gravity Waves

General Relativity predicts that every movement of matter produces gravitational waves. For example, when you wave your hand, you produce such waves. These, however, are tiny and undetectable. In contrast, when two black holes collide and merge to form a larger black hole, the gravitational waves can be substantial and should be observable.

In 1974, Russell Hulse and Joseph Taylor found indirect proof of the existence of gravity waves. They studied the rotation rate of binary pulsars, two neutron stars orbiting around each other. They found that the orbit is gradually shrinking by about 3.1 mm per orbit because of gravitational wave emission, exactly as predicted by General Relativity. This will cause the two stars to merge in another 300 million years. In 1993, Hulse and Taylor were awarded the Nobel Prize in Physics for this discovery.

Gravity waves have not yet been detected directly. Their discovery is a goal of a current experiment known as LIGO, the Laser Interferometer Gravitational Wave Observatory. The upcoming Advanced LIGO is expected to have the sensitivity to see gravity waves from black hole mergers. To date, our knowledge of distant space has come to us in the form of electromagnetic waves: visible light, infrared, radio, x-ray, gamma ray, ultraviolet, and microwave. In the future, studies of gravitational waves will open another window into the Universe.

## Fun with Black Holes

### Black Hole Evaporation

Stephen Hawking's Area Theorems prove that black holes can only increase in size—with one exception. It was Hawking himself who found the loophole. He realized that black holes can evaporate via particle production. Now known as "Hawking radiation," this effect implies that black holes do not live forever. Everywhere in the Universe, quantum fluctuations of spacetime (temporary fluctuations in the amount of energy predicted by quantum mechanics) cause pairs of particles and antiparticles to pop in and out of existence. They are very short lived and annihilate almost as soon as they are created. Such quantum fluctuations in the vicinity of a black hole can have interesting consequences. If a particle-antiparticle pair is produced just outside the event horizon, then the pair can split up. One of the particles escapes the black hole, while the other falls inside. To conserve energy, one can think of the one falling into the black hole as having negative energy. As a consequence, the mass of the black hole decreases. Because of this process the black hole can get smaller and smaller until eventually it evaporates completely.

The outgoing particles from the black hole appear as radiation with an effective temperature inversely proportional to the object's mass. In other words, as the black hole gets smaller and smaller, the temperature grows larger. Unfortunately, there are currently no direct observations of Hawking radiation from black holes. A solar-mass black hole radiates at a temperature of $10^{-7}$ K, negligible compared to the 3 K radiation of the CMB. In fact, solar-mass black holes are absorbing microwave background radiation and are getting bigger rather than smaller as a result. In contrast, tiny black holes would be much hotter than the microwave background and would emit radiation rather than absorbing it. Those with masses less than $10^{15}$ grams (a million billion times smaller than the Sun) would have already evaporated by now and hence are no longer present.

### Wormholes and Time Travel

According to General Relativity, two widely separated regions of space can be connected by a wormhole. Such a situation could arise if each region contained a black hole, with the two black holes sharing the same interior. In principle, alien races from these two worlds could encounter one another by entering inside the black hole, but unfortunately, it is not clear whether they could ever reemerge. In principle such a "traversable wormhole" could be used to pass from one apparently disconnected region of spacetime to the next. Science fiction television shows such as *Star Trek* frequently make use of this concept. No known classical processes could create wormholes. Quantum fluctuations might conceivably produce them—but only on the tiny scales where quantum gravity becomes important. The expected size of one of these would be the Planck scale, or $10^{-33}$ centimeters, not very conducive to human travel. If a mechanism could be found to make a traversable wormhole, then time travel would also become possible. You could in principle travel backward in time to kill your ancestors, but then how could you exist? This conundrum, known as the grandfather paradox, leads physicists to believe that time travel into the past is impossible.

### Are Black Holes the Dark Matter?

We've seen that black holes can be divided into three types: supermassive black holes at the centers of galaxies, solar-mass black holes at the endpoint of stellar evolution, and primordial black holes from the early Universe. The black holes in the first two categories definitely exist. Every galaxy has a supermassive black hole at its center, including Sagittarius A* in the Milky Way. Heavy stars that run out of nuclear fuel definitely collapse down to solar-mass black holes. Both these types of black holes form out of atomic matter. Yet Big Bang nucleosynthesis and the CMB restrict the amount of atomic matter to constitute only 5% of the total content of the Universe. Thus these two types of astrophysical black holes, constituted of atomic matter, cannot provide the dark matter in galaxies.

In contrast, primordial black holes are viable dark matter candidates. They were produced while the Universe was still in its infancy and could still exist inside galaxies today. The possible mass range for these black holes is between the mass of the moon and that of Earth. Of the existing candidates for dark matter, I would place primordial black holes as the third-most plausible. Weakly Interacting Massive Particles (WIMPs) and axions are more strongly motivated, as discussed later in this chapter. The existence of primordial black holes is hard to rule out but also difficult to prove. They do not emit signals of any kind and are difficult to detect. There is no experimental evidence for them. In some sense, primordial black holes make ideal dark matter candidates, because no one can argue with a speculative object that is impossible to see. However, such a situation is obviously quite unsettling, and so most cosmologists are pursuing other options.

### The Particle Zoo

In graduate school at the University of Chicago, I was introduced to the field of particle astrophysics. In this modern view, dark matter is thought to consist of some new kind of fundamental particles—not neutrons, protons, quarks, or any other constituents of ordinary matter, but something entirely new and different. I worked with my PhD advisor David Schramm, head of the Astronomy and Astrophysics Department, and with Michael Turner (Figure 5.3), at that time a rising star in cosmology. Their work was seminal in this fledgling field. They argued strongly for a Universe with a flat geometry consisting primarily of non-atomic constituents. They wrote important papers on Big Bang nucleosynthesis that showed definitively that atomic matter cannot constitute more than 10% of the critical density corresponding to a flat Universe. At the time, many astronomers still believed in a spherical Universe consisting only of ordinary matter.

In contrast, Schramm and Turner argued that only a flat geometry made any sense. Any other choice was unstable. On theoretical grounds, a spherical Universe should have recollapsed to a Big Crunch within $10^{-43}$ seconds after the Big Bang. Alan Guth's inflationary cosmology, which features an early exponential growth period of the Universe, drove the Universe to become flat. The combination of bounds from nucleosynthesis together with the preference for a flat geometry led to a Universe with a dominant component not made of atoms. Consequently Schramm and Turner were early believers in particle dark matter. As a graduate student, I came to believe that the solution to the dark matter problem relies on connections between the smallest subatomic particles and the largest galaxies, clusters, and superclusters.

FIGURE 5.3 (Left) Pierre Sikivie founded the field of axion searches. (Right) Michael Turner has been an early force in particle astrophysics. *(Left) Pierre Sikivie; (right) Department of Physics, University of Chicago.*

A plethora of possibilities exist for dark matter from the particle physics world: WIMPs, axions, sterile neutrinos, hidden sector dark matter, WIMPZILLAs, mirror matter, and others with equally whimsical names. The remainder of this chapter introduces some of the most interesting candidates for dark matter.

Luckily, as cosmologists, we don't need to invent new particles just for the purpose of explaining dark matter. The two best-motivated candidates from the particle physics point of view are WIMPs and axions. These are taken the most seriously because they automatically exist in particle theories for reasons entirely disconnected from the dark matter problem. They kill two birds with one stone. As we'll see, axions automatically arise in a proposed solution to a problem in quantum chromodynamics (QCD). WIMPs naturally come out of the theory if supersymmetry is the solution to a variety of problems in particle physics. WIMPs also appear in some theories invoking extra dimensions. These connections between particle physics and astrophysics are discussed further below. From the particle physics perspective, the motivation for the other candidates is not as strong as for axions or WIMPs, but they are still fascinating objects.

### Axions

Axions are hypothetical particles that come out of modern variants of QCD, the theory of strong interactions. The strong force holds together protons and neutrons inside the nuclei of atoms. Without it, atoms would disintegrate because of the electric repulsion between protons. All the protons in a nucleus have positive electric charges and would fly apart if they were not held together by the strong force. Yet quantum chromodynamics suffers from the *strong charge-parity (CP) problem*. CP symmetry states that the laws of physics should be the same if particles were switched with their antiparticles (C symmetry), and then left and right were transposed (P symmetry). The theory of QCD predicts a violation of this symmetry that is not observed. An elegant solution to the strong CP problem was proposed by Roberto Peccei and Helen Quinn. As realized independently by Steven Weinberg and Frank Wilczek, this solution automatically comes with new particles—axions. These particles are very light, slow, and abundant. Here on Earth there could be as many as a quadrillion in a coffee cup. In fact, if they exist, they are so plentiful they could solve the dark matter problem.

Soon after axions were proposed, cosmologists looked for evidence of their existence in astrophysical settings. A variety of arguments quickly ruled out the simplest models and placed strong constraints on the allowed masses. If axions were too light, there would be too many of them (the lighter the axions, the larger their numbers). With their enormous total mass density, they would have driven the Universe to rapidly collapse into a Big Crunch. The Cosmos would never have achieved its present age of 13.8 billion years. In contrast, if axions were too heavy, they would destroy red giant stars and supernovae. Created in the centers of these stars, axions would drain the energy out of them as these particles left the stars. The allowed mass window for axions ranges from a billionth to a quadrillionth of the mass of a proton.

In 1983 Pierre Sikivie (see Figure 5.3) had a clever idea for axion detection. In the presence of magnetic fields, axions would convert into photons. Experimentalists certainly know how to detect light. The Axion Dark Matter Experiment at the University of Washington in Seattle is using Sikivie's idea to look for axions. This experiment consists of a microwave cavity inside a large superconducting magnet. The cavity acts like a tuning fork that resonates with a frequency comparable to the proposed axion mass and enhances the conversion to photons. In another experiment, the CERN Axion Solar Telescope in Geneva is searching for photons from the Sun created by axion conversion in the magnetic field of the solar neighborhood. Axions have not yet been discovered, but some of the most interesting axion masses and interactions will soon be probed.

### WIMPZILLAs, Mirror Matter, and Q-Balls

Other dark matter candidates from the particle world, although less strongly motivated than the WIMP and axion favorites, provide interesting alternatives. One intriguing possibility is the WIMPZILLA particle.[18] Like the

WIMPs of the next section, these are particles that interact only with gravity and the weak force. Yet the WIMPZILLAS are much heavier, with each one weighing as much as a quadrillion protons. These particles could have been produced in a nonequilibrium stage of the Universe, such as at the end of an exponentially expanding early inflationary era.

Many other dark matter candidates exist as well, including mirror (or shadow) matter. Some scientists hypothesize that the Universe is symmetric (unchanged) under mirror reflection. Then a *mirror particle* exists for every ordinary particle. These mirror species can have widely differing masses from their ordinary counterparts and may not interact much with them. Other than playing a role in the gravitational dynamics of galaxies, mirror matter is hidden. If these particles accumulated into stellar-sized objects, they would have been detected in MACHO searches, but they have not been spotted.[19]

Another possibility is that dark matter consists of Q-balls or other *nontopological solitons*. These objects can be thought of as clusters of particles confined to little balls that are energetically stable against evaporation into free particles. The Q-ball state minimizes energy while keeping a new type of charge (Q, associated with a new symmetry) constant. Although mathematically fascinating, in my opinion these are not the most compelling candidates for dark matter.

### The WIMP Miracle

In the view of most physicists, the best candidates for dark matter are Weakly Interacting Massive Particles (WIMPs). The name carries the most important information about these particles. "Massive" refers to their relatively large mass, which is roughly 10 to 1,000 times as much as that of the proton. "Weakly interacting" suggests that their only interactions besides gravity are by means of the weak force. If the theories are right, billions of WIMPs could be passing through our bodies every second (but we would never notice).

There are four fundamental forces of nature: gravity, electromagnetism, the strong force, and the weak force. Gravity keeps us standing on the surface of Earth, keeps planets moving around the Sun, and holds together all the mass in the Galaxy. The definition of "matter" is that it feels the attractive force of gravity, and clearly dark matter falls in this category. Electromagnetism causes the attractive force between electrons and protons and keeps magnets attached to refrigerators. The strong force holds together the neutrons and protons inside atoms. We know that dark matter does not interact via the strong or electromagnetic forces. Strongly interacting dark matter would have been seen in underground or gamma-ray detectors, or would have destroyed the Galactic halo and Earth. Electrically charged dark matter also would have been detected in cosmic ray and gamma-ray experiments. In addition, because it breaks molecular bonds, it would have degraded electronic components of man-made satellites.[20]

That leaves the fourth force: the weak force, which is responsible for some types of radioactivity as well as the fusion process that fuels the Sun. As its name suggests, the weak force is far weaker than the strong or electromagnetic forces. Although dark matter is immune to strong and electromagnetic forces, weakly interacting dark matter is a very interesting and well-motivated proposition.

A major reason that WIMPs are taken so seriously as dark matter candidates is a feature known as the WIMP miracle. Because these particles undergo only weak interactions (in addition to feeling gravity), they miraculously turn out to have the right abundance today to solve the dark matter problem. In the early days of the Universe, WIMPs roamed the Universe as part of the primordial soup. They frequently collided with other particles. Many WIMPs are their own antiparticles, and pairs of them annihilated in this hot dense environment. The weak force determined the strength of this annihilation process. Then, as the Universe cooled down and expanded, the interactions became less commonplace. WIMPs no longer ran into one another, and the annihilation stopped. Their residual numbers are determined by the weak interactions that controlled the annihilation. The nice feature of this theory is that the WIMPs remaining today automatically have the right abundance to solve the dark matter problem. This coincidence, based entirely on the weak force felt by the particles, is a major reason for the interest in WIMPs as dark matter.

Supersymmetry

Further rationale for the interest in WIMPs is that they solve several problems at once. They naturally come out of particle theories proposed for reasons entirely unrelated to the dark matter problem. WIMPs arise naturally in supersymmetry (SUSY for short), a mathematically elegant extension of the Standard Model of particle physics. Before the advent of SUSY, it was impossible to combine spacetime symmetries with the forces of nature into a single theory. SUSY provides connections between bosons and fermions, particles with integer and half-integer values of spin (the concept of spin is described further in the discussion of the Higgs particle in Chapter 6). It provides a mechanism for adding gravity to the Standard Model. It predicts high-energy unification of the electromagnetic, strong, and weak interactions in one force.

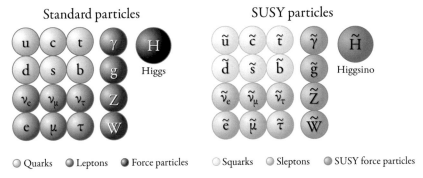

Standard particles                    SUSY particles

○ Quarks  ● Leptons  ● Force particles    ○ Squarks  ○ Sleptons  ● SUSY force particles

**FIGURE 5.4** The supersymmetric (SUSY) extension of the Standard Model doubles the numbers of particles in the Universe. Every particle has a SUSY partner, such as the squark ($\tilde{q}$) for the quark, the photino ($\tilde{\gamma}$) for the photon, and the Higgsino for the Higgs. The tilde over the top of the letter indicates the SUSY partner. The lightest SUSY particle, probably a combination of Higgsino and SUSY force particles, is the dark matter candidate.

One of the major motivations for building the Large Hadron Collider, the gigantic atom smasher at CERN in Geneva, is the search for SUSY. I once asked Nobel Prize winner David Gross what he hoped would be discovered at the Large Hadron Collider, and he said, "SUSY, SUSY, SUSY!" Part of his interest in SUSY is its role in string theory, a proposed "Theory of Everything" in which strings replace particles as the fundamental constituents of nature (as discussed later in the chapter).

SUSY doubles the numbers of particles in nature. If SUSY is right, then every particle we know about has a supersymmetric partner (Figure 5.4). The photon's partner is the photino; the partners of the quarks are called squarks; and the W particle has a partner called a Wino (pronounced "weeno," as opposed to the name given to someone who overindulges in wine). All SUSY particles undergo the same weak interactions as their ordinary partners. Yet SUSY particles are heavier than their counterparts.

The lightest supersymmetric particle (LSP) would be a good WIMP dark matter candidate. All SUSY particles heavier than the LSP are short lived and would decay via a chain of reactions to ever lighter ones until they reach the LSP. This particle is stable against further decay. The best bet for the LSP would be a SUSY particle called the neutralino, which is a mixture of a Higgsino (the partner of the Higgs), a photino (partner of the photon), and a wino (partner of the W). Neutralinos are so named because they are electrically neutral, a requirement for any dark matter candidate. They are also their own antiparticles and would naturally annihilate among themselves in the early Universe, just as required for the WIMP miracle that predicts the correct dark matter abundance today.

The stability of the LSP is due to a property of SUSY theories known as R-parity. All standard model particles have an R-parity of +1, whereas SUSY particles have an R-parity of −1. The overall value of R-parity must be conserved in every interaction, including in the decay chain of heavier SUSY particles into lighter ones. Because R-parity must be conserved, the LSP can't possibly decay into ordinary matter of the Standard Model: a particle with an R-parity of −1 can't convert to a particle with parity +1. This new parity property guarantees the existence of a long-lived SUSY particle that could be responsible for dark matter.

## Extra Dimensions

A second home for WIMPs in modern particle theory is extra dimensions. The number of spatial dimensions may be larger than the three we ordinarily experience. In the 1920s, Theodor Kaluza and Oskar Klein first realized that the existence of additional dimensions would enable the unification of gravity with the other forces of nature. The modern variant of this idea is string theory, an attempt to unify all four forces of nature into a single Theory of Everything, as mentioned above. In this theory, strings are the fundamental constituents of the Universe, and the particles we are familiar with are vibrations of these strings. String theory only works if the Universe has 10 dimensions, where 6 of these are new spatial dimensions. Because we only experience 3 spatial dimensions, the rest must be so tiny that we are not aware of them.

New particles are associated with these extra dimensions. Some models predict a new conserved quantity called "Kaluza-Klein parity," named after the two men who first proposed force unification in extra dimensions. Similar to particles with R-parity in SUSY, there are additional particles in the theory with Kaluza-Klein parity. The heavier Kaluza-Klein particles decay to ever lighter ones until they reach the one with the lowest mass. This lightest Kaluza-Klein particle is stable against any further decay and makes a good WIMP dark matter candidate.[21]

## WIMPs in the Human Body and a Tennis Match

Recently I was on an episode of the television show, *Through the Wormhole with Morgan Freeman*. The episode was called "Is Nothing Really Something?" and it aired in July 2012. Filming it was a really fun experience. We went to a tennis court in Topanga Canyon, California, in the middle of December. I played tennis against myself. I was dressed in black as "dark matter Katie," and I competed against myself dressed in white as "regular matter Katie." I'm not sure who won. I can say it was freezing and I had to take my winter coat on and off in between takes.

Anthony (Tony) Lund, the science writer for the episode, asked me how often a WIMP would collide with a nucleus in a human. I told him I wasn't sure, but that I did know that billions of these particles pass through our bodies every second. My guess was one collision per century, and I joked about the WIMP death theory: when a WIMP hits you, you die (just kidding!). After that, Chris Savage (my former student) and I did the calculation. We computed the numbers of WIMP interactions with the different types of nuclei in the human body and found that the most frequent collisions are with oxygen and hydrogen. The results really surprised me. For a 150-pound human, the number of hits could be as frequent as one every minute! For heavier WIMPs, the maximum number would be somewhat lower—one every month (the experimental bounds on the interactions of heavier WIMPs are stronger). Either way, the count rate of WIMP interactions with humans is much higher than I would have imagined.

Next Tony asked us whether these interactions were dangerous. My initial response was that they couldn't possibly be, because weak interactions are so "weak." This time I was right. Although WIMPs are a source of radiation, it is negligible compared to the danger from cosmic rays raining down from the Galaxy through Earth's atmosphere. Cosmic rays are a potentially serious health hazard for pregnant women or airline personnel. In contrast, WIMPs are harmless.

WIMPs are great dark matter candidates. Other than gravity, their only interactions are by means of the weak force. Because this force is so weak, detecting these particles is a challenge. Typically they move right through almost anything, even the entire Earth. They interact very infrequently, and deposit only a small amount of energy when they do. Nonetheless, detectors are currently searching for the signatures of such scattering events. Chapters 7 and 8 describe the experimental efforts to discover these particles and the tantalizing hints that a signal may have been found.

# SIX

# The Discovery of the Higgs Boson

It looks like the Higgs;
It smells like the Higgs;
It tastes like the Higgs.
It must be the Higgs.

## Atom Smashers: The Large Hadron Collider

The Large Hadron Collider (LHC) is the highest-energy particle accelerator on Earth and is one of the most amazing structures ever built. To quote the British newspaper *The Guardian,* it is "the biggest machine in the world." The LHC accelerates protons to move at up to 99.9999993% of the speed of light. The primary science goals are to discover the Higgs particle that gives other particles their masses; to search for supersymmetry (SUSY); to solve the dark matter problem; and to seek out the unknown—new forces, new particles, or new surprises.

The most exciting physics development at the LHC has been the discovery of the famous Higgs particle, the important missing ingredient in the Standard Model of particle physics. Its existence has been predicted for decades as an explanation for the masses of particles in the Universe.[1] This chapter discusses the importance of the Higgs boson to particle physics as well as the experimental route to its discovery.

## CERN

The LHC is located in the countryside on the outskirts of Geneva, at the European Organization for Nuclear Research (CERN). Roughly 10,000 physicists and engineers work there, coming from 111 countries, including the United States. Geneva is a nice place to live, with ski resorts only an hour away, wonderful fondue (melted cheese or melted chocolate), great food and espresso in the CERN cafeteria, and an excellent view of the Jura Mountains.

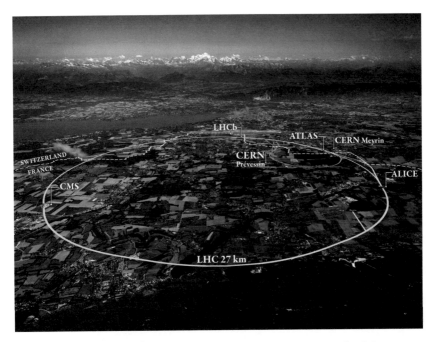

**FIGURE 6.1** The Large Hadron Collider at CERN near Geneva, Switzerland. Protons are accelerated in opposite directions around the ring (drawn in on the surface of Earth) in an underground tunnel 27 kilometers (17 miles) around. CERN scientists work at two main sites: CERN Meyrin and CERN Prévessin. The particles collide at the four intersection points where the CMS, ATLAS, LHCb, and ALICE detectors are located. *CERN.*

The LHC accelerates particles around a circular ring that is roughly 27 kilometers (17 miles) in circumference, crossing the border of Switzerland and France in four different places. In the 1990s, the preeminent accelerator was at Fermilab, an hour west of Chicago. Experimentalists were able to jog the 4 miles around the farmland directly above the Fermilab ring for exercise. Jogging around the ring at CERN would constitute two-thirds of a marathon, yet even for the best runners is impossible. They would have to cross the Franco-Swiss border repeatedly as well as traverse fences between neighboring farms. Driving across the diameter of the LHC ring from one of the major particle detectors to the other takes 40 minutes on winding roads through farmland with cows and barns (Figure 6.1).

At about 12 feet in diameter, the LHC's concrete-lined underground tunnel is just about wide enough for a small car. Inside, roughly 300 feet beneath the surface, two adjacent parallel pipes carry proton beams along a circular path in opposite directions. The protons are accelerated around the ring in a vacuum (the air pressure in the pipes is lower than on the moon). Once the two beams are moving fast enough, physicists in the control room focus them with pinpoint accuracy and smash them into each other at four different points along the ring, where detectors known as CMS, ATLAS, LHCb, and ALICE are located. Emerging from the collisions are sprays of new particles, which the detectors identify and study. From these collisions, physicists hope to discover new particles, new properties of forces, and new fundamental physics.

A wonderful video on YouTube captures the exciting physics at CERN. Type the words "LHC rap" into any search engine, and the song written and recorded by science journalist Kate McAlpine (aka Alpinekat) will come up.[2] The scientific content of the video is dead on. It helps explain the basic concepts of the experiments: "The LHC accelerates the proton and the lead, and the things that it discovers will rock you in the head."

### The Accelerator

CERN accelerates protons in stages up to almost the speed of light. The protons start out slow and are successively sped up in auxiliary accelerators. First the protons enter a linear particle accelerator, the LINAC 2, where magnets focus the beam along a single straight line and accelerate them to about a third of the speed of light. Next the protons enter a series of circular accelerators (the Proton Synchrotron Booster, the Proton Synchrotron, and the Super Proton Synchrotron), and finally they enter the main LHC accelerator ring itself. Once inside the ring, the protons' paths are controlled by superconducting magnets with magnetic fields of 8 Tesla, roughly 200,000 times the magnetic field of Earth, or 2,000 times stronger than a refrigerator magnet. These magnets weigh more than 27 tons. Superconducting magnets can carry large currents with virtually no resistance; for the magnetic coils to remain superconducting, they operate at the extremely low temperature of 1.9 K ($-456°$ F). Particles with positive and negative electric charge curve in opposite directions in a magnetic field. These magnets steer the beams in their opposing circular orbits as they are accelerated by electric fields to nearly the speed of light.

Once the protons are inside the main LHC ring, they keep circulating for roughly 15 hours. One orbit takes only 90 microseconds (90 millionths of a second), amounting to 10,000 revolutions every second. The proton beam consists of bunches of protons (containing 100 billion protons per bunch) rather than one continuous beam. At any one time there are about 1,300 bunches in the ring.

When the protons reach the desired energies, physicists in the LHC control room use magnets to squeeze the beams to induce collisions. At an interaction point, the 100 billion protons in a bunch are focused to a size of 60 microns, the width of a human hair. These bunches collide with one another at intervals of 25 nanoseconds. When proton bunches from the two opposing beams cross, roughly 20 proton-proton collisions take place. The 20 collisions per beam crossing that occurs every 25 nanoseconds add up to almost a billion collisions every second. Very few of these produce interesting new physics; most simply create backgrounds that obscure the important events. Sophisticated computer modeling is needed to isolate the data that are worth recording. Most of the protons miss one another in any given beam crossing and continue to circle around the ring.

After the protons have been cycling for 15 hours or so, the magnetic fields are ramped down, so that the process can be started all over again. The residual proton beam has so much energy that it could be extremely destructive. It must be removed from the ring and steered into a separate cavern housing a "beam dump," where it is completely absorbed and stopped. The beam dump consists of absorbers made of 7-meter-long carbon cylinders contained inside steel cylinders. Its core is water cooled and surrounded by 800 tons of concrete and iron shielding. The proton beam ends its life there.

## The Detectors

The two proton beams racing in opposite directions around the LHC ring smash into each other at four different intersection points. Each of these points is at the heart of a giant detector that measures the sprays of new particles created in the collisions. Two of the detectors are multipurpose instruments whose primary goals are the hunt for new fundamental physics: CMS (Compact Muon Solenoid) and ATLAS (A Toroidal LHC Apparatus). The other two, ALICE and LHCb, have been designed to study more specific physics.

The CMS and ATLAS detectors are enormous—as high as five-story buildings. Thousands of physicists and engineers have worked for decades to build these huge structures. The ATLAS detector is roughly 150 feet long, 40 feet in diameter, and weighs 15 million pounds (about the same as the Eiffel Tower). The CMS detector has smaller dimensions (it is about 70 feet long) but weighs nearly twice as much because of its high iron content—that is why it is called "compact." Both detectors have the same goal of capturing as much information as possible about the particles produced in the collisions as they move outward from the interaction point. Figure 6.2 shows Fabiola Gianotti, the spokesperson for the ATLAS team, with the detector.

Both ATLAS and CMS have similar designs with an onion-skin structure. They consist of three concentric stages of detector elements that study particles moving outward from the collision point. These are known as a central parti-

FIGURE 6.2 (A color version of this figure is included in the insert following page 82.) Fabiola Gianotti, spokesperson for the ATLAS team of 3,000 scientists at CERN, made the first public announcement of the discovery of a new particle, most likely the Higgs boson. *CERN.*

cle tracker, calorimeters (that measure particle energies), and, farthest out from the interaction region, the Muon Spectrometer. The purpose of this design is to reconstruct the properties of the particles produced in the interaction. For example, if a Higgs particle were created in a proton-proton collision, it would quickly decay to other particles, which would further decay, and so on. The three stages in the detector would measure the tracks and energies of all the particles in this decay chain, starting from the collision point and moving outward. Using computers to combine all the data, scientists can reconstruct the entire event and look for evidence of the Higgs particle.

In the ATLAS detector (Figure 6.3), the detector element closest to the central interaction point is referred to as the Inner Detector. This tracks the paths of the innermost particles produced in the proton-proton collisions. A powerful 2-Tesla magnetic field (50,000 times the strength of Earth's magnetic field) bends charged particles into curved paths and helps scientists identify their electric charge and speed. Particles of positive and negative electric charge curve in opposite directions in a magnetic field, and the degree of curvature depends on how fast they are moving. The Inner Detector extends from the central interaction point out to a radius of about 3.5 feet. Because of their proximity to the proton-proton collisions, both the detector and the

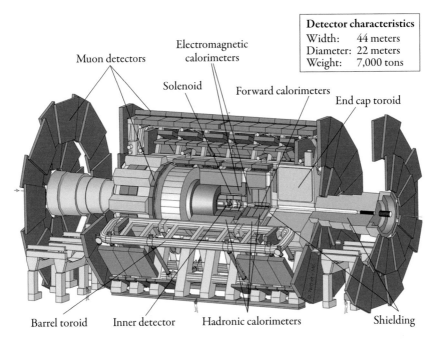

Muon detectors

Electromagnetic
calorimeters

Solenoid

Forward calorimeters

End cap toroid

**Detector characteristics**
Width:      44 meters
Diameter: 22 meters
Weight:    7,000 tons

Barrel toroid    Inner detector    Hadronic calorimeters    Shielding

FIGURE 6.3 The ATLAS detector at CERN. The two proton beams come in from the left and right sides through the beam pipe and collide in the heart of the detector. The outgoing particles from the interaction point traverse a series of concentric stages of detector elements. *ATLAS Experiment © 2013 CERN.*

electronics reading out the data have to be able to withstand a great deal of radiation.

Just outside the Inner Detector is a set of calorimeters, that is, detector elements that measure the energies of the particles moving outward from the interaction. This information helps scientists reconstruct the masses and energies of the original particles created in the collision. As they pass through the calorimeters, particles lose energy, and most of them stop. Only muons and neutrinos penetrate through to the third detector stage, the muon spectrometer. Here the properties of the muons are measured. Neutrinos escape the detector entirely without leaving any signal whatsoever. By combining the information from all three stages of detector elements, scientists can identify the chain of particles emerging from the collisions of two proton beams.

The CMS detector has a concentric onionskin design similar to ATLAS. An inner tracker tracks charged particles and measures their speeds. Next are calorimeters to determine particle masses and energies.[3] Farthest out from the collision point are the muon detectors. Russian collaborators on the CMS detector contributed brass casings of decommissioned artillery shells of the Russian Northern fleet. These were melted down and then machined in Minsk as an endcap for the calorimeter.

In addition to the general-purpose detectors ATLAS and CMS, there are two other main detectors, ALICE and LHCb, located at the two other interaction regions in underground caverns. Researchers designed these detectors for specific physics goals. Roughly 10% of the time, the protons in the beam pipes at the LHC are replaced by lead ions. The ALICE detector (A Large Ion Collider Experiment) studies the collisions of these heavy lead ions. When lead ions smash into each other, they produce a hot mixture of subatomic quarks and gluons, known as the quark-gluon plasma, that replicates the primordial soup of the early Universe. Today quarks are confined in complex subatomic particles, such as protons or neutrons. However, at high energies or temperatures, the strong force that holds the quarks together is much less effective. The binding force was so weak in the early Universe that the quarks existed as independent entities, isolated from one another. The ALICE detector is able to study the behavior of these free quarks in a setting that mimics the high temperatures that prevailed soon after the Big Bang. A detailed understanding of the origin of the confinement of elementary particles in the protons and neutrons of our Universe today is an important subject of inquiry for the ALICE detector. Without the transition from free quarks after the Big Bang to more complex particles today, we would not exist.

The fourth major detector, LHCb (LHC-beauty), is searching for charge-parity symmetry (CP) violation. CP violation in the early Universe is thought to play a crucial role in producing the matter-antimatter asymmetry—the enormous excess in the abundance of particles over antiparticles in the Universe today. Cosmologists assume that there were equal amounts created after the Big Bang. Yet, as discussed in Chapter 5, in today's Universe the number of particles exceeds the number of their antipartners by a ratio of a billion. The origin of this asymmetry is not understood. Theorists believe that CP violation was essential in generating more particles than antiparticles. For example, the decays of superheavy particles in GUT baryogenesis models discussed in Chapter 5 are CP violating. LHCb experimentalists hope to find clues to solving this puzzle. Any new sources of CP violation discovered at CERN would require exciting new physics beyond the Standard Model.

There was a great deal of excitement when the LHC first started firing protons around the ring on September 10, 2008. Unfortunately, only a week later, a serious explosion forced the collider to shut down. A short-circuit in a faulty electrical connection between two magnets caused mechanical damage, and 6,000 kilograms of liquid helium coolant leaked into the accelerator tunnel. The coolant loss caused 50 of the powerful superconducting magnets to "quench": they heated up, lost their superconducting properties, and their magnetic fields collapsed. The resulting pressure shock wave from the quench was powerful enough to rip the enormous magnets from their attachment to the concrete floor and lift them around the tunnel. It took more than a year to repair the damage.

On March 30, 2010, the LHC started recording physics data again. In fact, to make up for lost time, it continued running through winter 2011, when it would usually be turned off, as it is more expensive to run in winter. To be on the safe side, the accelerator restarted below its maximum energy capacity, but after a shutdown and upgrade scheduled into 2015, the LHC's energy capacity comes close to its original design specifications. Prior to the shutdown, the collision energy for proton-proton interactions was 8 TeV (TeV stands for a trillion electron volts, roughly 1,000 times the energy equivalent of the mass of a proton). After the upgrade, the LHC energy of 13 TeV, close to the original design plan for the accelerator, allows for exciting searches for unknown new physics.

### Discovery of the Higgs

The existence of the Higgs particle has been predicted since 1964 as a necessary ingredient in the Standard Model of particle physics to explain the origin of particle mass. It is also known as the Higgs boson, or simply, the Higgs. The word "boson" refers to the spin of the particle; bosons are particles with an integer value of spin $(0, 1, 2, \ldots)$. In contrast, fermions have half-integer values of spin $(+1/2, -1/2, +3/2, -3/2, \ldots)$. Spin is a quantum number characterizing different types of particle behavior; fermions, for example, obey the Pauli exclusion principle discussed in Chapter 5 in the context of white dwarfs (whereas bosons do not). Spin can be thought of by analogy to a spinning top with a discrete variety of possible rotation speeds. The Higgs has zero spin, unlike quarks, which have spin $+1/2$ or $-1/2$.

Although our current knowledge of particles and their interactions is sophisticated, many deeply puzzling questions remain. One is the origin of the wide variety of masses of particles in the Universe. Protons weigh roughly 1 billion electron volts (1 GeV, equivalent to $10^{-24}$ grams), electrons weigh 0.5 million electron volts (0.5 MeV), pions weigh 135 MeV, and so forth. Particle physicists do have an explanation for the origin of the mass of the proton, which constitutes most of the ordinary matter in the Universe. The proton mass can be understood by quantum chromodynamics (QCD), the theory of strong interactions. Protons consist of up quarks, down quarks, and the gluons that hold them together. QCD tells us that the energy of any isolated quark is infinitely large. To avoid this untenable situation, a sea of quarks and antiquarks springs into existence around any lone quark to cancel out this infinite energy. In the words of Nobel laureate Frank Wilczek, each of the constituent quarks in the proton is accompanied by a "big color thundercloud" of energy that produces what we experience as the mass of the proton.[4] Frank developed these concepts while he was a 21-year-old graduate student of David Gross at Princeton. In 2004 Gross and Wilczek jointly received the Nobel Prize in Physics with David Politzer, who independently had the same basic ideas. However, the origin of masses of the individual elementary particles in the Universe—including isolated quarks and gluons—is not explained by QCD.

Thus we have the Higgs boson. It is named after Peter Higgs, one of six physicists whose theories predicted the existence of the Higgs particle in 1964; the others were Robert Brout, François Englert, Gerald Guralnik, Carl Richard Hagen, and Thomas (Tom) Kibble. Peter Higgs is shown in front of the CMS detector at CERN in Figure 6.4.

According to their theory, the Higgs particle is responsible for the creation of mass by means of the *Higgs mechanism*. At the very high temperatures of the early Universe, all particles were massless. Then as the temperature dropped, the Universe experienced what is known as a phase transition—a change similar to the formation of ice as water cools. In the process, the change in the Higgs field led to masses for the particles. The term "field" refers to the energy that exists throughout space; the Higgs particle accompanies this field. The Higgs gives particles their masses in two different ways.

The Higgs boson imparts mass to quarks, electrons, muons, and many other elementary particles by slowing them down. A useful analogy would be to imagine weaving a path through a crowded room of dancing people. If the people weren't there, you could run straight through—much like a massless particle moving unimpeded at the speed of light. But the presence of the dancers impedes your movement, like molasses. This is the equivalent of a particle moving through the field of the Higgs boson. The more sluggish the particle becomes as it encounters the Higgs field, the heavier it will be. Many fundamental particles obtain their masses in this way.

The Higgs imparts mass to the W and Z bosons in a slightly different way. These particles are responsible for the weak interactions, the least powerful of the four forces of nature. The Higgs mechanism that gives mass to the W and Z bosons is a little more complicated. At high temperatures above the phase

FIGURE 6.4 (A color version of this figure is included in the insert following page 82.) (Top) The CMS detector at CERN during construction. (Bottom) Peter Higgs, who won the 2013 Nobel Prize in Physics for predicting the existence of the Higgs boson, in front of the CMS detector. *CMS and CERN. Copyright © 2008 CERN.*

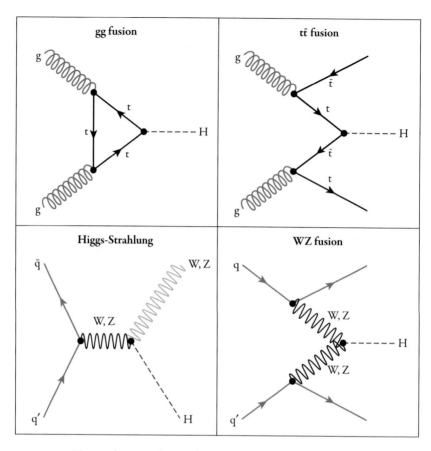

**FIGURE 6.5** The mechanisms for production of Higgs (H) particles at CERN. The quarks (q) and gluons (g) inside the protons collide to make Higgs particles. The dominant mechanism is the gluon-gluon fusion displayed in the top-left panel: a top (t) quark loop allows the conversion of two gluons to a Higgs particle. In these diagrams, overbars indicate antiparticles. The primes over the quarks indicate that two different types of incoming quarks may be involved (for example, an up quark from one of the colliding protons and a down quark from the other).

transition, all particles were massless. After the transition, the Higgs field split into four components: the particle that is now the Higgs boson plus three other massless particles. These three combined with (or in particle physics parlance, "were eaten by") the W and Z bosons; this combination produced the massive W and Z bosons that drive the weak force in today's Universe.

Even with the discovery of the Higgs boson, particle physicists still cannot explain why particles have the variety of different masses they do, but at least the Higgs explains why they have masses at all. The Higgs particle is so impor-

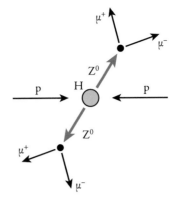

**FIGURE 6.6** Collisions of two protons can create a Higgs particle, which then decays into two $Z^0$ particles. Each of the $Z^0$ bosons converts to two muons ($\mu^+\mu^-$ as shown) or two electrons. The resulting four particles can be detected in CMS or ATLAS, and the mass of the Higgs can then be reconstructed.

tant that Nobel Prize winner Leon Lederman's 1993 book dubbed it "The God Particle," a name that (although unpopular with the physics community) has stuck in the popular literature.[5] The search for the hypothesized Higgs boson has been on for decades.

Figure 6.5 shows theoretical predictions for the four primary mechanisms for Higgs production at the LHC. In collisions of two protons, it is the constituent quarks and gluons inside the protons that smash into each other. This process gives rise to the creation of new particles. For example, two gluons can exchange a top or bottom quark and in the process, known as gluon-gluon fusion, create a Higgs particle. This mechanism is the predominant one for Higgs production at the LHC and is shown in the top-left panel of the figure.

Three other mechanisms for Higgs production are shown in the figure, including the fusion of a top quark with a top antiquark known as $t\bar{t}$ fusion (illustrated in the top-right panel). After a Higgs is created, it doesn't last very long before it decays into other particles, particularly pairs of Ws, $Z^0$s, photons ($\gamma$s), bottom quarks, and taus. For example, the Higgs can decay to two $Z^0$ bosons, each of which in turn decays to two electrons or two muons, giving rise to a "four-lepton" event, as shown in Figure 6.6. Or, it can decay to a pair of photons. The ATLAS and CMS detectors can observe all these final particles to compare to predictions with and without the Higgs production. The photon-photon ($\gamma\gamma$) and ZZ decay channels are especially important, because they allow the mass of the new particle to be measured with precision. Figure 6.7 shows an actual event recorded with the CMS detector at the LHC.

In the case where the Higgs decays to two high-energy photons, CMS or ATLAS can measure the photon energies and directions. The sum of the energies of the two photons is known as the *invariant mass*, or $m_{\gamma\gamma}$. Experimentalists can add up the number of events found in the detector that produced a

**FIGURE 6.7** (A color version of this figure is included in the insert following page 82.) A computer reconstruction of an actual proton-proton collision event using real data recorded with the CMS detector in 2012. The event shows evidence for the decay of a Higgs particle into a pair of photons (in the color version, dashed yellow lines leading to green towers). *CMS and CERN.*

given value of the invariant mass. Figure 6.8 plots the results from the data in the CMS detector from 2011 and 2012. Even by eye, one can see a bump at 125 GeV. The large number of events at this energy cannot be explained by background noise alone. This bump is undeniable evidence for the existence of a new particle that has a mass of 125 GeV.

On July 4, 2012, both the CMS and ATLAS teams announced that a new particle had been discovered.[6] To quote the ATLAS paper, the new particle is "compatible with the production and decay of the Standard Model Higgs boson." Although it was not yet official, most physicists immediately believed that the Higgs boson had indeed been found. Within a few months, even the most cautious experimentalists were ready to call it the discovery of the Higgs. The fundamental constituents of all atomic matter are quarks, leptons, and gauge bosons, and now the last missing ingredient has been found: the Higgs particle. The Standard Model of particle physics is now complete.

The bump at 125 GeV in Figure 6.8 not only proves that the Higgs exists but also may provide the first hint of physics beyond the Standard Model of particle physics. The first measurements of the bump were a bit too high to match standard predictions. Theorists, including postdoctoral fellow Nausheen Shah

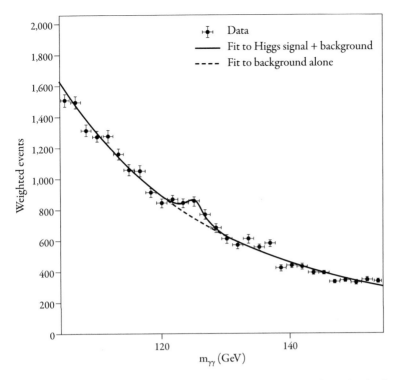

**FIGURE 6.8** The discovery of the Higgs particle: the bump in the data taken by the CMS detector at CERN is due to production and decay of a Higgs particle. The black points with error bars indicate the numbers of observed events from high-energy proton collisions into two photon final states, for a variety of values of the two-photon invariant mass $m_{\gamma\gamma}$ (the sum of the energies of the two photons). The solid line shows the fit to the data for Higgs signal plus background noise; the dashed line shows only the background. The bump in the data is caused by a newly discovered Higgs particle with a mass of 125 GeV (125 proton masses). *CMS and CERN.*

at the University of Michigan, are speculating that SUSY may be the origin of the excess—the same SUSY described in Chapter 5 as the possible origin of dark matter. Yet with the analysis of more data, the agreement with the Standard Model already looks better. Scientists will have to wait and see whether the discrepancy persists in future data.

The discovery of the Higgs is important for several reasons. First, it solves the problem of mass in the Standard Model of particle physics. The existence of quarks and leptons as the fundamental constituents of matter had been confirmed for decades, yet without a Higgs, these particles would be massless and consequently always moving at the speed of light. If the Higgs were not found, it would have created a difficult but interesting situation for theorists, who would have had to think long and hard about alternatives. Second, the Higgs was the missing ingredient in the unification of the electromagnetic and weak forces into the *electroweak force*.

Third, the Higgs is the first scalar, or spin-zero, particle that has ever been found. As mentioned earlier in this chapter, all particles have spin: quarks have spins of $+1/2$ or $-1/2$, whereas photons have spin 1. Although the existence of spin-zero particles had been postulated for many reasons, including the Higgs and the inflaton (responsible for the early accelerated expansion of the Universe), none had previously been found. Their existence is a conceptually important leap in particle physics.

The discovery of the Higgs is cause for jubilation in the world of physics. The euphoria among physicists worldwide has been tremendous. *Science* magazine hailed the discovery of the Higgs boson as "the top science breakthrough of 2012."[7] The Nobel Prize in Physics 2013 was awarded jointly to François Englert and Peter Higgs "for the theoretical discovery of a mechanism that contributes to our understanding of the origin of mass of subatomic particles, and which recently was confirmed through the discovery of the predicted fundamental particle, by the ATLAS and CMS experiments at CERN's Large Hadron Collider."[8]

# SEVEN

# The Experimental Hunt for Dark Matter Particles

What has been the most important contribution to society from the high-energy physics done at particle accelerators? Ten billion dollars have been spent at CERN, and I think it is valid for nonscientists to inquire what this investment has done for them. When I ask people this question, some look puzzled, as they can't imagine much direct value. Others reply that new technology always leads to new spinoffs, and studies show that each research dollar reaps billions in returns—although those who give this answer typically cannot name any particular spinoffs from high-energy experiments. The most common answer is the human thirst for understanding our world, its nature, and its contents. This drive for exploration has also enabled us to control our environment and thrive as a species.

I would agree with these answers, but they are not my primary motivation. For myself I would have to say that when I play with mathematics or physics to solve mysteries of the Universe, I get excited. It makes my brain tingle. It can be better than any cocktail. The beauty of a painting is in the mind's response to combinations of wavelengths of light; the beauty of music in the response to compression waves of sound; the beauty of physics equations also registers in the mind in a different way. Sadly, the beauty of mathematics is accessible to a smaller number of people. Imagine trying to appreciate a poem written in Swedish without having learned the language—it would be just a bunch of unrelated sounds. The same is true for physics: without having learned the language of mathematics, it can look like a string of meaningless symbols and its beauty is hard to access.

As to the question of the societal value of particle physics done at CERN, there is actually one clear but surprising answer, a spinoff that has changed our daily lives forever. I'll save it for later in the chapter. For now, let's turn to the hunt for the dark matter in the Universe, the majority of the mass in the cosmos. This chapter outlines the three approaches scientists take in the experimental hunt for dark matter, and the next chapter tells the stories of the detectors currently taking data and their exciting results.

## The Three Prongs of the Hunt for Dark Matter

Experimental cosmologists are searching for dark matter particles in the Universe with three different approaches described in detail in this chapter (Figure 7.1). One method is to study the collisions of high-energy proton beams at the gigantic atom smasher at CERN. Sprays of particles produced by the collisions there could include the missing dark matter. Second, dark matter particles from the Galaxy might interact with atomic nuclei in direct detection experiments located in underground laboratories beneath mountains or in abandoned mines. Finally, dark matter particles might annihilate among themselves in distant parts of the Galaxy; products of the annihilation might be seen in indirect detection experiments on satellites or under the ice at the South Pole. These three types of experiments all rely on the same basic physics: Weakly Interacting Massive Particles (WIMPs) interact with ordinary matter by means of the least powerful of the four fundamental forces, the weak force—but not the electromagnetic or strong forces. The three lines of attack described in this chapter—production in colliders, direct detection, and indirect detection—are complementary. Together they should be able to solve the dark matter problem.

## Dark Matter at the Atom Smasher at CERN: Missing Energy plus Jets

Now that the Higgs boson has been discovered, the other major science goal of the LHC is to discover the identity of dark matter. Dark matter particles might be produced in proton collisions at the LHC and could be identified in a variety of ways. One clear signature relies on a combination of two effects: *missing transverse momentum* plus a number of *jets*.

Any dark matter particles produced at the LHC interact so weakly with matter that they escape right out of the detector. They pass through the three stages of detector elements without making tracks or depositing energy. As a result, experimentalists will observe a momentum imbalance in the observed particles. The particles that can be measured in an event will appear to violate momentum conservation. Scientists will conclude that some unseen particle must have carried the missing momentum out of the detector. This search for the violation of

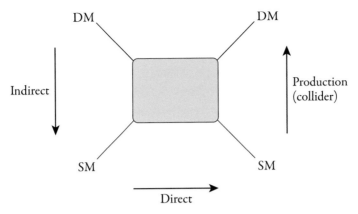

**FIGURE 7.1** One diagram illustrates all three prongs of the experimental hunt for dark matter particles. Weakly Interacting Massive Particles (WIMPs), labeled by the letters DM, interact with ordinary matter particles, labeled SM (for the Standard Model of particle physics), via the weak interactions symbolized by the box in the middle of the diagram. The box can represent a variety of mechanisms, including intermediary W bosons, Z bosons, or Higgs bosons. The three arrows indicate the three different detection approaches. The upward-pointing arrow on the right side of the figure illustrates WIMP production in atom smashers such as at CERN in Geneva. In this case, the diagram is read from bottom to top: two protons (labeled SM at the bottom of the diagram) collide with one another to produce outgoing WIMPs (labeled DM at the top of the diagram). Detectors at CERN search for the WIMPs created by this interaction. Alternatively, the arrow that points to the right on the bottom of the diagram illustrates the basic idea of laboratory direct detection experiments. In this case, the diagram is read from left to right. Here a WIMP from the Milky Way Galaxy (DM on the upper left) hits a nucleus in the detector (SM on the lower right), scatters off of it, and deposits a detectable amount of energy. In the end, the WIMP (DM on the upper right) leaves the detector again. The nucleus (SM on the lower right) has gained energy from the interaction, and this energy can be measured. Finally, the downward pointing arrow on the left indicates indirect detection, where the annihilation of WIMPs in space produces energetic photons, neutrinos, and positrons (all from the Standard Model of particle physics) that may be found in detectors here on Earth. In this third detection method, the diagram is read from top to bottom.

momentum conservation is similar to the way neutrinos were discovered. Measurements of beta decay seemed to violate energy conservation, unless a new unseen particle, in this case the neutrino, was carrying energy out of the detector. At the LHC, energy conservation cannot directly be tested. It is impossible to measure the total energies of the particles in an event, because a lot of energy is lost down the beam pipe (where it is impossible to place a detector). Yet the component of the particles' momentum perpendicular to the beam, known as the *transverse momentum,* can be identified. When dark matter particles fly out of the detector, the transverse momentum will show an imbalance.

A second signature of dark matter particles produced at the LHC will be one, two, four, or more *jets* in the detector. When two protons collide at the LHC, they produce a chain of particles that includes quarks. Because of quark confinement, quarks cannot exist as isolated entities. Instead they hadronize—they bind with quark and gluon companions to compose more complex objects like neutrons or protons. As a result, instead of a single particle track, each quark produces a collimated spray of particles known as a jet. The chain of particles resulting in dark matter production should produce accompanying jets of particles.

Thus scientists can identify dark matter production by observing jets and missing transverse momentum. With these two pieces of information, sophisticated computer programs can be used to reconstruct the entire decay chain. The goal is to identify the existence of a dark matter particle and then to measure its mass and its interaction with matter.

We've seen that one of the best-motivated possibilities for WIMP dark matter arises from SUSY. As described in Chapter 5, if SUSY correctly describes nature, then every particle in the Standard Model of particle physics has an additional SUSY partner. For example, photons would have photinos, quarks would have squarks, and electrons would have selectrons. The lightest of these SUSY particles could provide dark matter for the Universe.

The proton-proton collisions at the LHC might create pairs of SUSY particles, specifically, squark-squark, gluino-gluino, or squark-gluino pairs. An example is seen in Figure 7.2. These pairs would decay almost immediately to a series of ever lighter SUSY particles until, finally, they reach the lowest-mass SUSY particle, the dark matter particle. This would escape the detector without depositing any energy; it would apparently vanish. The signature of such a SUSY event would then be missing momentum due to the disappearance of the dark matter particle together with one or more jets from the quarks in the decay chain.

For decades, the search for such SUSY events has yielded no discovery and has pushed the mass of the hypothetical SUSY particles to higher and higher values. But scientists are optimistic that SUSY could be discovered at the LHC, because it operates at such high energies.

As yet the LHC has found no events indicative of SUSY. Low-mass SUSY particles cannot exist, or they would already have produced signals in the detectors. The current lower limits on some SUSY masses are approaching TeV

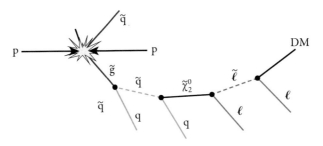

FIGURE 7.2 Signatures of dark matter particles at the Large Hadron Collider will be missing energy plus jets. Because dark matter escapes the detector without interacting, there will be an imbalance in the measured momenta of the final particles in the direction perpendicular to the beam. The quarks and leptons emerging from the decay chain will also produce observable jets of particles, a second signature of dark matter. In the figure the dark matter consists of the lightest supersymmetric particle, symbolized by DM. Tildes over the letters indicate SUSY particles, and the letter $\ell$ refers to leptons, that is, electrons, muons, or tau particles.

scales—the natural scale expected if SUSY is to play a role in weak interaction physics and in solving the dark matter problem. (Again, TeV stands for a trillion electron volts, an energy equivalent to about 1,000 times the mass of a proton.) If the LHC does not find evidence of SUSY, then a future higher-energy accelerator will be required for conclusive tests of the theory; unfortunately, funding such a venture will prove to be difficult.

The motivation for WIMPs goes beyond SUSY. Regardless of whether SUSY proves to be a correct description of nature, the WIMP miracle is based on the idea of GeV to TeV mass particles governed by weak interactions and gravity. This basic idea of a dark matter particle motivated by electroweak physics is a very strong contender for explaining the bulk of the mass in our Universe.

Even if new fundamental physics such as supersymmetry is discovered at the LHC, we still won't know whether the dark matter has really been found. All we will know is that the unknown particle lasts long enough to escape the detector (milliseconds); we still won't know whether it lives as long as the age of the Universe (14 billion years). Dark matter particles have to be stable on this long time scale, something the LHC will never be able to test. Confirmation from complementary astrophysical experiments will be required to prove that the dark matter particle has really been discovered.

## What Has CERN Done for Society?

I began this chapter by asking a question: What has been the most important contribution to society from the particle physics done at accelerators like the LHC? I suggested that the innate human desire to understand our world is a big part of the answer. Yet, in addition, a surprising spinoff of the research has changed the world.

There's an amazingly practical answer to the question of what particle experiments have done for society: the creation of "www" that we type into our computers every day. Timothy (Tim) Berners-Lee was working as a computer scientist at CERN when he invented the World Wide Web. Physicists from all over the world come to CERN to run their experiments but must return to various institutions in different countries to teach their courses and rejoin the people in their lives. Wherever they are, they need to be able to access the data coming from the accelerator in Switzerland. Berners-Lee had the brilliant idea to post the data stream on a computer web that allowed these physicists to obtain whatever they needed online. In 1990 he implemented the first successful communication between HTTP (hypertext transfer protocol) client and server via the internet. His contribution is as important to society as was the Gutenberg Bible produced in the 1450s by the first modern printing press. For this work, Tim Berners-Lee was knighted by Queen Elizabeth in 2004.

CERN continues to play a major role in the field of computer science. The LHC produces an enormous amount of data. It stores several petabytes of information per year, where 1 petabyte corresponds to 1 million gigabytes, or roughly 1,000 trillion bytes. For comparison, a typical desktop computer can store up to 1,000 gigabytes. The gigantic amount of data currently accumulating is handled by the LHC Computing Grid, a network that distributes the data on multiple computers throughout the world.[1] It is the biggest computer science project on Earth.

Further, to quote CERN's François Flückiger, director of the CERN School of Computing and inductee into the Internet Hall of Fame: "In 1991, 80% of the internet capacity in Europe for international traffic was installed at CERN, in building 513! It is no surprise that the performance of Tim Berners-Lee's first web server impressed the world: it was at the heart of the European internet, just a few hops from most destinations."[2]

The LHC produces so much data that most of it has to be discarded. Only a small subset can be stored for later study. A great deal of thought has been put into the decision of what to save. Here theorists play a major role. The expectations for what science could come out of the experiments drives not only the design of the detectors but also the choices of which incoming data to save. "Cuts" are made on the data. Before an event is stored, it has to meet several criteria: it must have a predetermined amount of energy in particular detector elements, a large enough number of outgoing particle tracks, and so forth. Only events surviving these cuts are investigated further. Most of the incoming data is just dumped. It is frightening to think that great discoveries may be missed as a consequence. The physics may be so foreign to our current understanding that we can't possibly guess that the relevant data would be worth saving and studying.

### Will the LHC Bring Doomsday?

Before the LHC turned on, some people outside the scientific community predicted doomsday scenarios. If the proton-proton collisions had the right energy, they could produce mini–black holes. Some claimed that these black holes could grow in size and rapidly engulf Earth (see the imaginative video on the website http://www.youtube.com/watch?v=BXzugu39pKM). A high school teacher filed a lawsuit in federal court in Hawaii attempting to stop the accelerator from turning on. Among the safety studies carried out by CERN were discussions of this issue. The scientists concluded that such a scenario was not going to happen. In fact, in my opinion, the overblown press given to this notion amounted to bad reporting. First, the cosmic rays bombarding us daily through the atmosphere attain higher energies than anything the LHC can create. Second, even if tiny black holes formed, they would evaporate immediately because of Hawking radiation (see Chapter 5). Of course, quantum mechanics leaves open the possibility, however remote, that anything can happen with nonzero (though negligible) probability. For example, there is a finite probability that you will pop out of existence before you finish reading this paragraph, but I wouldn't worry. To quote Nima Arkani-Hamed of the Institute for Advanced Study, there is some minuscule probability that "the Large Hadron Collider might make dragons that might eat us up."[3]

The *Daily Show* with Jon Stewart aired an incredibly funny episode pitting world-famous particle theorist John Ellis at CERN against the high school teacher who filed the lawsuit to stop the LHC.[4] Now that the LHC has been up and running for a few years, it's clear that it's not going to create black holes that will swallow Earth.

### Direct Detection: Abandoned Mines, Alpine Tunnels, and Nightclubs in Jerusalem

Complementary to the creation of dark matter in accelerators is the detection of astrophysical dark matter particles. These particles traverse the halo of our

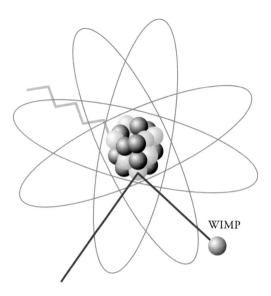

WIMP

**FIGURE 7.3** Direct detection experiments: a Weakly Interacting Massive Particle (WIMP) from the Galaxy scatters off of a nucleus in a detector, leading to a nuclear recoil and small amount of energy deposit that can be measured.

Galaxy with velocities of about 250 kilometers per second (560,000 miles per hour). Because they do not feel electromagnetic or strong nuclear forces, the experiments to directly detect these particles have to be extremely sensitive to weak interactions, the least powerful of the four forces.

Occasionally, when a WIMP passes through a detector here on Earth, it interacts with one of the nuclei in the detector. The collision of a WIMP from our Galaxy with a nucleus is much like the collision of two billiard balls (Figure 7.3). The nucleus recoils and heats up. The goal of direct detection experiments is to measure the tiny amount of energy transferred to the nucleus by the WIMP. Because the predicted count rates are extremely small—less than one count per kilogram of detector material per day—these experiments are difficult. Some experiments currently taking data consist of more than 100 kilograms of detector material, and future plans are to build detectors with 10 metric tons of material (1 metric ton is 1,000 kilograms)

I started working on WIMPs while I was a postdoctoral fellow at Harvard. In December 1984, I attended a one-week program in theoretical physics at the Hebrew University in Jerusalem. The program was designed to be educational for young cosmologists. It included talks by the eminent lectur-

FIGURE 7.4 Andrzej Drukier (left) and Leo Stodolsky (right) wrote the first paper (published in 1985) on detecting particles with weak interactions via their scattering off of nuclei. Their paper was geared toward looking for neutrinos. At the time Leo was director at the Max Planck Institute for Theoretical Physics in Munich, where Andrzej was a visiting scientist. *(Left) Andrzej Drukier.*

ers James Gunn, Alan Guth, Stephen Hawking, Michael Turner, and Steven Weinberg. At a New Year's Eve party in Jerusalem, toward the end of the conference, I met Andrzej Drukier, one of the participants of the program. A brilliant but eccentric Polish physicist, he speaks simultaneously (and sometimes unintelligibly) in a mixture of French, Polish, English, and a smattering of German. In Israel, the 31st of December is not the usual date for celebrating New Year's. A group of us from the conference were happy to follow Andrzej to the Cinemateque, a movie theater that had been converted into a dance club for the evening. As we danced and shared whiskey and champagne, Andrzej told me about a paper he had coauthored with Leo Stodolsky at the Max Planck Institute in Munich on ideas for detecting cosmic neutrinos. These particles, much like WIMPs, engage only in weak interactions with nuclei. I was intrigued by what seemed at the time to be a radical proposal. When we got back to the United States, I followed up with him. Their paper laid the foundation for the idea of building detectors for particles with weak interactions (Figure 7.4). *"My Dinner with Andrzej"* steered my research interests toward dark matter.

In January 1985 I visited Rockefeller University in Manhattan for 2 months. Andrzej met me in the city, and we took the train to Princeton in New Jersey to talk to Edward (Ed) Witten, a Fields medalist (the equivalent of a Nobel Prize in mathematics) at Princeton University. There are few women in physics now, but in the 1980s there were even fewer. Recently one of my friends who is a professor at Princeton told me that when Andrzej and I walked into Ed's office and closed the door, it caused a bit of stir. People wanted to know what we were talking about. We were discussing WIMPs!

Ed and his graduate student Mark Goodman applied the ideas for neutrino detection proposed by Drukier and Stodolsky to WIMPs instead. The physics is the same, because neutrinos and WIMPs are all weakly interacting particles, but because WIMPs are much heavier, their anticipated count rates in detectors would be higher. Goodman and Witten worked out rough estimates for the numbers of WIMP interactions that could be detected and showed that these detectors were worth building. Ed is thought by many to be one of the most brilliant physicists alive. He is a leading figure in string theory. Many young theorists closely follow whatever he is working on as a guide for their own research. Occasionally he turns his mind to topics outside string theory. The paper with Mark on the subject of WIMP detection had a tremendous impact.[5]

In the spring of 1986 we arranged for Andrzej to visit us at the Harvard / Smithsonian Center for Astrophysics for a semester. There I introduced him to Harvard graduate student David Spergel (Figure 7.5). The three of us sat around a table in a small meeting room at the Center and came up with ideas about aspects of dark matter detection. Our discussions and calculations played a pioneering role in the field of dark matter phenomenology, the connection between dark matter theory and experiment. This work proved to be a catalyst for both David's and my careers. After graduate school, David moved to Princeton, where he has remained ever since. He later became a MacArthur fellow, led the cosmology analysis of the cosmic microwave background data gathered by the Wilkinson Microwave Anisotropy Probe (WMAP) satellite (discussed in Chapter 3), and is currently chair of the Astronomy Department at Princeton. The first thing he did with his MacArthur Fellowship money was buy a foosball table. David and I had both been Princeton undergrads, and the tradition in our time for math, physics, and astronomy students was to play foosball (instead of video games). In fact another physics major, Louis Salkind, competed in the national foosball championships after graduating. Later he became one of the first quants, devising mathematical models for Wall Street. As one of the original members of the financial firm D. E. Shaw in New York, Louis always kept a foosball table in his office.

**FIGURE 7.5** David Spergel (left, at the time a graduate student at Harvard University) and the author (right; at the time a graduate student at the University of Chicago). In 1985, Andrzej Drukier, Spergel, and I performed calculations that developed the nascent field of dark matter phenomenology, the connection between dark matter theory and experiment. *(Left) David Spergel.*

Drukier, Spergel, and I computed the predicted rates for WIMP scattering off of nuclei in a variety of possible detector target materials. We took into account the distribution of WIMPs in the Milky Way as well as the detailed particle physics of WIMP interactions. The count rates we predicted were high enough to inspire experimentalists to start developing the required detector technology.

As it turns out, our original numbers for predicted detection rates were high. The first types of WIMP candidate we considered were photinos and other simple SUSY particles (recall that photinos are the SUSY partners of photons; see Chapter 5). By chance, these simple particles we first considered led to predictions of unusually large count rates compared to other WIMPs. Without intending it, we initially happened upon the most optimistic cases for dark matter detection. We inadvertently started with the tip of the iceberg and didn't realize that most dark matter candidates remained beyond the reach of existing detector technology.

In retrospect, we were lucky. Experimentalists might have been discouraged from starting the hunt for WIMPs if they had realized that the particles we originally considered were not representative of typical WIMPs. The funding agencies might have balked if the difficulties had been apparent. Instead, the race to build the detectors began immediately. Scientists thought it could be done based on our early numbers. The first detectors reported no signal. Instead of discovering WIMPs, they placed upper limits on WIMP scattering rates with nuclei. Photinos, with the highest predicted count rates, were quickly ruled out as dark matter candidates. The absence of a signal implied that the WIMP-nucleus interactions had to be weaker than our early estimates. Although some dark matter particles were ruled out, most WIMPs had scattering rates below the sensitivity of the early detectors.

Nowadays both the theoretical understanding and the experimental sensitivity have improved tremendously. Whereas we did our initial estimates using pen and paper (I even computed Feynman diagrams by hand), later researchers wrote computer codes to automate the calculations.[6] The computer programs are quite complex: they consider all possible SUSY dark matter candidates; they include all relevant constraints from high-energy experiments; and for any given WIMP candidate, they make predictions for both direct and indirect detection experiments. The earliest code, named DARKSUSY, is regularly updated and is still used by theorists and experimentalists alike.[7] Other scientists use MicrOMEGAs, a competing computer program for dark matter calculations.

Twenty-five years after the first efforts, the detector technology has improved to the point where much lower count rates are accessible. The possible range of WIMP candidates—the full iceberg—is being whittled down by the experiments. Direct detection dark matter searches are ongoing worldwide, including in the United States, Canada, Germany, France, England, Switzerland, Italy, Japan, China, and Korea (Figure 7.6).

## Particle Physics of Direct Detection

To estimate the viability of direct detection in the laboratory, we had to take into account two different types of physics. The rates of WIMP interactions in detectors depend on both the particle physics of the WIMP interaction and the astrophysics of the WIMP distribution in the Galactic halo.

On the particle physics side, a variety of WIMP-nucleus interactions are possible. The two most common are known as spin-independent (SI) and spin-dependent (SD) scattering. These two types of interactions have different origins and would produce different scattering rates. In a spin-independent collision of a WIMP with a nucleus, the WIMP interacts with each neutron and proton in the nucleus, and the individual contributions add up to enhance the

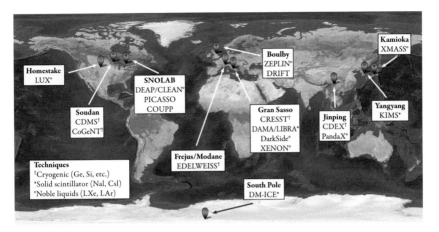

**FIGURE 7.6** Underground laboratories (indicated in boldface) are the sites of direct detection dark matter experiments worldwide (experiments listed at their respective locations). These hunt for dark matter by searching for the energy deposited when a WIMP strikes a nucleus in the detector. The key indicates the three types of techniques used in the experiments: cryogenic experiments operating at low temperatures measure the heat deposited by the interaction; solid scintillators absorb the energy and reemit flashes of light; and noble liquids produce scintillation light as well as electrons that later convert to photons. *Michael Woods and Mani Tripathi, University of California, Davis.*

overall number of collisions with the nucleus as a whole. The rate is amplified by a *coherence factor* that scales as $A^2$, that is with the square of the atomic mass A (the total number of neutrons plus protons). For the case of nuclei lighter than the WIMP, the enhancement factor is even larger, scaling as A to the fourth power ($A^4$). Thus heavier nuclei make far more effective WIMP targets. To maximize the number of WIMP scatters with SI interactions, experimentalists use massive elements in their detectors. Chapter 8 discusses in detail detectors made of a variety of materials. For example, we will see that the CDMS and CoGeNT detectors are made of germanium, an element whose predominant isotope has an atomic mass of $A = 74$, much larger than hydrogen's $A = 1$. The WIMP spin-independent scattering rate in germanium detectors can be 30 million ($74^4$) times as large as it would be with a hydrogen target.

In contrast, for spin-dependent scattering, a WIMP interacts only with the total spin of the nucleus. The coherence factor $A^2$ is lost. Although individual neutrons and protons inside the nucleus have spins of plus or minus 1/2, typically nuclei have an even number of nucleons, and the total spin of the nucleus adds to zero. In this case there are no spin-dependent interactions, and the WIMPs just pass right through. But there are exceptional nuclei with one extra (unpaired) neutron or one extra proton. These nuclei have nonzero spin and do have spin-dependent interactions with WIMPs. An effective target for spin-dependent scattering is hydrogen, which is made of one unpaired proton with spin +1/2. Other heavier elements also have isotopes with nonzero spin. Sensitivity to spin-dependent interactions requires the use of targets with unpaired neutrons or protons.

To compute WIMP-nucleus spin-dependent scattering rates in detectors, I had to learn how to perform nuclear shell model calculations. Ironically, I had been taught the subject a few years earlier, while I was a graduate student at Columbia University in New York—but I hadn't bothered to learn it. The professor of the class, Gary Feinberg, was a nuclear physicist who had worked in this area himself. He always came to class wearing a bow tie and was a very caring professor. Sadly, I had decided the subject was uninteresting, and so I ignored it. Unfortunately for me, it was a significant part of the final exam, and I was in trouble! I suspect my grade defined passing in the class, as Feinberg didn't want to flunk me out of graduate school. My unwillingness to learn the subject at the time was especially ironic, because the model was first developed by the only female theoretical physicist to ever win a Nobel Prize, Maria Goeppert Mayer. How funny that these same calculations, which nearly cost me a career in physics, became important for my work in dark matter! The basic idea of the nuclear shell model is that neutrons and protons in nuclei can be treated as occupying a series of concentric shells. They pair up, yet in some cases, there is an extra neutron or proton that conveys overall spin to the nucleus. It is with this excess spin that a WIMP can interact. I easily learned how to use the nuclear shell model to estimate count rates in a variety of different detector materials.

The relative importance of these two types of scattering—spin independent versus spin dependent—is determined by the exact nature of the WIMP. Even in the context of SUSY WIMPs, there are different candidates that have predominantly one type of interaction or the other. Thus, to search for as many WIMP types as possible, experimentalists often design a detector to have isotopes of elements with both paired and unpaired nuclei that are sensitive to spin-independent and spin-dependent scattering, respectively. For example, in the CDMS experiment, Ge-74 (spinless) is the dominant constituent of the detector, whereas Ge-73 (spin 9/2) makes up 7.73% of natural germanium, so that the detector can search for both spin-independent and spin-dependent scatters of WIMPs.

## Astrophysics of Direct Detection

The astrophysics of the Galaxy also impacts WIMP counts in detectors. Drukier, Spergel, and I were the first to take into account the distribution of

WIMPs in the Milky Way in these calculations; this distribution is an important ingredient in predictions for detectors. We treated the dark matter as a single smooth component known as the Standard Halo Model. Whereas the atomic matter in the Galaxy has electromagnetic and strong interactions that cause it to settle into a flattened disk, the dark matter is essentially collisionless (weak interactions occur very rarely among WIMPs in the galaxy). Thus the dark matter component remains in a large diffuse halo with its shape and structure determined only by gravity.

The Standard Halo Model treats the Milky Way as being essentially spherical.[8] The dark matter halo behaves like a gaseous sphere characterized by a "temperature" set by the typical speeds of the particles—about 250 kilometers per second (560,000 miles per hour). The dark matter particles move in a variety of orbits around the center of the Galaxy. Most of the orbits are not perfectly circular and can instead be elliptical or even chaotic. Some extend out to large radii before returning to the center. The count rates in detectors are the highest for the slowest WIMPs and drop exponentially for more energetic particles. The rate falls to zero for particles faster than about 540 kilometers per second (120,000 miles per hour)—the Galaxy's *escape velocity*. At these speeds, WIMPs are no longer bound to the Milky Way: they move quickly enough to escape the gravitational attraction of the Galaxy.

An important factor in these calculations is the *density profile*: the dependence of the number of dark matter particles on the distance from the Galactic Center. It has been known for decades that the WIMP density peaks at the center and then falls off roughly as the square of the distance from the center. Computer simulations of galaxy formation of better and better resolution since the 1980s have obtained improved predictions for this quantity. In the hierarchical clustering model, small objects form first and then merge to make ever larger structures. According to the simulations with the highest resolution, all of these astrophysical objects, on scales ranging from Earth-sized to galaxies and clusters, are found to have the same density profile (known as the Navarro, Frenk, and White profile).[9] The simulations predict that the WIMP density peaks at the center of the dark matter halo and then falls off initially as the distance from the center and then farther out as the distance to the third power. Though the inner profile is uncertain, the basic premise of a dark matter density peak at the Galactic Center is clearly correct.

Computer simulations of galaxy formation show that the simplest Standard Halo Model is imperfect. First, the smooth component of the dark matter may have a different, more complicated, distribution than the one used in standard calculations. Second, the halo is likely to be clumpy rather than smooth, with streams and tidal debris of matter. Our Galaxy formed from mergers of smaller objects; in fact, some of these mergers are still ongoing inside the Milky Way. We do not have enough information about the details of the Milky Way to know its precise dark matter structure, although observations of stellar streams and motions are starting to improve our level of understanding. The uncertainties in the halo models affect the predicted detection rates and make comparison among results of different detectors more difficult to interpret. For the discussion here, we stick with the Standard Halo Model but will return to these important subtleties later on.

The key issue for direct detection experiments is the number of WIMPs in the detectors here on Earth. The Sun is located about 24,000 light-years from the Galactic Center. Here in the solar neighborhood, the WIMP density is thought to be roughly 0.4 GeV per cubic centimeter, which amounts to about one WIMP per coffee cup (for a WIMP mass of 100 GeV). The Sun is moving around the center of the Galaxy at a speed of about 250 kilometers per second. As a consequence, we are moving into what appears to be a wind of WIMPs. The relative velocity between the WIMPs and us is what determines the count rate in detectors.

Drukier, Spergel, and I noticed an important effect. Because Earth orbits around the Sun, the relative speed between Earth and the WIMP wind changes with the seasons. Thus the count rate in detectors should record differing numbers of impacts at different times of the year. The WIMP signal should peak in June and have a minimum in December. None of the competing spurious effects (such as radioactive contamination) are expected to have this yearly variation. Such an annual modulation has been seen in a number of detectors and is the driving force behind recent claims of WIMP discovery. We will return to this subject in Chapter 8.

## False Signals

The biggest problem for all direct detection experiments is background noise, which can mimic a WIMP interaction. Cosmic rays are a problem. These energetic charged particles roam the Galaxy, buffeted by magnetic fields. Some of them enter Earth's atmosphere, where they decay into muons and neutrinos that relentlessly bombard us with radiation. Indeed, cosmic rays are the worst occupational hazard of being a flight attendant or an airline pilot. The numbers of these particles are higher the farther up you go in the atmosphere. Because the magnetic field of the Earth draws them preferentially toward the poles, flights from the United States to Europe that travel over the North Pole can be quite dangerous. The Sun experiences an 11-year cycle in solar activity.

During the peak years of the cycle, solar flares and winds push large numbers of cosmic rays toward Earth. These can disrupt communications satellites and wreak other havoc. To avoid damage to the fetus, pregnant women would be well advised to check the National Oceanic and Atmospheric Organization webpage for a live webstream of the daily cosmic ray count and to avoid flying on high-radiation days.[10]

To eliminate spurious signals from cosmic rays, WIMP detection experiments have to be placed deep below the surface of Earth in abandoned mines or in tunnels beneath mountains. The cosmic rays get trapped in the ground, and most cannot reach the underground caverns. The deeper the mine, the better is the background reduction. Most people associate mines with tragedies in West Virginia or Pennsylvania, but underground facilities can also be the sites of scientific experiments. An example is the Homestake Mine in South Dakota, an abandoned gold mine with a cavern 4,850 feet (almost a mile) below the surface of the Earth. Experiments are currently taking data in other deep underground sites around the world. One of the chief expenses associated with keeping these mines operational is that they naturally fill with water, which needs to be pumped out constantly to keep the caverns functional.

Once most cosmic rays are blocked by going deep underground, the major source of contamination is radioactivity from the detector itself or from the surrounding rock. Whether it is radon naturally occurring everywhere in the ground, problematic radioactivity from the experimental apparatus itself, or radioactive isotopes in the detector material, this radioactive background is exceedingly hard to remove. Scientists try to make accurate estimates of the false signal arising from various sources, but it is always difficult to know whether all the backgrounds have been properly accounted for. These background estimates are always a major source of disagreement and discussion among competing groups at scientific meetings. The backgrounds must be studied very carefully before a claim of WIMP detection can be taken seriously.

### Early Experiments

Soon after we finished our early theory paper, Andrzej approached David and me about participating in a project he had discussed with experimentalist Frank Avignone. Frank's high-purity germanium detectors that were already taking data for other purposes could be used as dark matter detectors.[11] They suggested that we work together to analyze existing data from these experiments and look for WIMP signals. Frank is a lot of fun: he's an Italian American from New York who was in the Navy in World War II and calls himself "an old seadog." I last saw him at the Silver Jubilee of Dark Matter Detection Meeting in Washington State in June 2012, where we had quite a bit of really good local white wine. Turning 80 in 2013, he is still charming and going strong. He has more money coming in from government grants than most young researchers garner. When it comes to low-background-noise experiments, he is the best. When his wife died, he was fixed up on a date and as he says, when you want to get intimate with someone, you ought to marry her. So now he has remarried. He is a true scholar and a gentleman (and an officer too). At one of the UCLA Dark Matter Meetings held in Marina del Rey, California, a few years ago, I went to a nightclub with him on Santa Monica Boulevard, and we danced up a storm. He told me that when I remarry, he wants to walk me down the aisle, as my own father is dead. Now he says I have to hurry up. We shall see.

Frank Avignone and his colleague Ron Brodzinski had the best low-background high-purity germanium detectors. Together with Andrzej Drukier, they wanted to see what limits they could place on dark matter, given the data they already had. Because the project required the help of theorists, Graciela Gelmini and David Spergel joined the group. Because I was moving across the country to take up a new postdoctoral position in Santa Barbara, I had to drop out of the collaboration. Their paper showed that superlow-background-noise semiconductor spectrometers could be used to put significant bounds on DM candidates.[12] As no signal was found in the detector, this paper produced the first experimental bounds on WIMPs. In June 2012, the Silver Jubilee Meeting at the Pacific Northwest National Laboratory in Washington State was held in memory of Ron Brodzinski and his role in this work.

Soon after, Bernard Sadoulet, an exceptional French experimentalist, moved from CERN to the University of California, Berkeley, to embark on a new career in dark matter detection. He is a dashing gentleman, a good friend, and a brilliant scientist. Bernard teamed up with David Caldwell at University of California, Santa Cruz, and Blas Cabrera at Stanford University to found the Cold Dark Matter Search (CDMS) experiment. They were important pioneers of direct detection experiments. These three succeeded in getting the U.S. research community started in investing in detectors designed specifically to look for dark matter. In some sense they created the field of dark matter detection as an experimental science. The first detection limits by CDMS were published in 1987. Sadoulet also founded the Berkeley Center for Particle Astrophysics. To this day, the CDMS experiment is a world leader in obtaining some of the best results in this field and in training some of the brightest young people. We'll return to CDMS in Chapter 8, when we discuss current experimental work.

The influential and creative experimentalist Peter Smith played a similar seminal role in England. He persuaded the British funding agencies to pursue dark matter searches in the Boulby Mine, formerly a potash mine, a kilometer from the Yorkshire coast. The resulting series of ZEPLIN experiments, using liquid xenon, made important contributions to dark matter searches.[13]

### Lifetime of a Dark Matter Project

Direct detection experiments have a wide variety of different designs, yet they share similarities in their evolution. The typical project has five different stages. The first is conceptualizing the experiment and its design. Next come laboratory studies to determine whether the idea is feasible. Stage three moves on to small-scale underground experiments. In stage four, the full-scale experiment is implemented and initial data are analyzed. Finally, the experiment is upgraded and data analysis is improved.

The first stage is the work of a few charismatic founders who propose a new experimental approach for WIMP hunting. The typical team is small—two or three senior scientists with a few students and technicians. This high-risk stage may cost about half a million dollars and lasts for 1 or 2 years. It is high risk because only about 10% of the ideas for new approaches survive to the next stage.

The second stage is fun. It grows the team to about 10 people, who spend 10–14 hours a day in a lab checking that the real device performance is compatible with the design simulations. This stage can cost up to a few million dollars and lasts a few years.

Stage three takes the experiment underground, where the background noise can be measured and the device can be fine-tuned. It is still the work of a small team. Some sites are enjoyable, such as the Italian Apennines, whereas others are murderous. Who wants to spend months at a time in an underground mine in a small town somewhere in South Dakota? The extreme is the case of indirect detection experiments at the South Pole. Nowadays there is a well-heated dormitory, but not long ago a temperature gradient existed even where people slept; lowering your legs to the ground en route to the restroom meant encountering really cold air. This stage also costs a few million dollars and lasts a year or so. Because modern simulations of backgrounds are incredibly precise, most experiments survive this stage and move on to the next one.

The full implementation stage is expensive, requiring 5–10 million dollars and the participation of many groups. A typical collaboration may involve 30–50 scientists and many different organizations, often from many countries.

The sociology of the group shifts. Although the leadership may still consist of the original founders, not all great physicists are good managers. Rumors abound about internal conflicts that hamper the progress of the experiments. Younger members of a group feel stifled and want to strike out on their own. As the chances of success grow, the pressure becomes enormous. Experiments at this point report either positive or negative results. Yet history shows that the statistics of the data at this stage are never sufficient. Most of the experiments require upgrades. Now that the techniques have been proven to work, experiments that are modular in design scale up by installing significantly more identical modules. There are always improvements, such as in the sensors and in the software, permitting better characterization of events and hence making it easier to reject background noise. This final stage is the longest, lasting a decade and costing millions of dollars a year. It is the bread-and-butter stage of an experiment, with many rewards: papers, conference presentations, PhD theses for students, and (one hopes) results. Several experiments are now in this stage and are releasing the exciting results described in Chapter 8.

What is to be expected for the future? Typically, about 10–15 years after conceptualization, it becomes clear that a certain class of detectors cannot be improved much more. Other competing experimental approaches enter the development stage. Theoretical predictions are pushed by experimental results into different regimes of WIMP masses and scattering rates. The existing detectors are no longer optimal and are phased out, and some of the original founders of the experiments begin to bow out. A new generation of scientists with new ideas and ambitions enters the arena. The current status of experiments is a mix of the old and the new, with both reporting interesting results.

### The New Era

Since the early experiments, the sensitivity to WIMP signals has improved by at least a factor of 1,000. It has taken tremendous creativity in the experimental realm to achieve such developments. The funding agencies of the United States as well as of Italy, Germany, France, China, Japan, Korea, and other countries are dedicating ever-larger sums of money to the experiments in the hopes of solving the dark matter problem. On the theoretical side, our early work opened the way for a deluge of important papers from astroparticle physicists, many of them my friends.[14] These are exciting times! The proposals and calculations we made 25 years ago have given birth to a rich worldwide search for dark matter and most recently to tantalizing hints of detection. Various unexplained anomalous results have emerged out of these experiments (see Chapter 8), and any of them may herald the discovery of dark matter.

### Indirect Detection: Annihilations in Space and at the South Pole

A third approach to WIMP searches is *indirect detection:* the hunt for astrophysical signals of WIMP annihilation products. Many classes of WIMPs are their own antiparticles and annihilate among themselves into a variety of lower-energy particles. The end states of the annihilation include positrons (antipartners of electrons), photons, and neutrinos. Scientists are searching for all three of these particles with detectors mounted on satellites in space, with strings of phototubes embedded in the ice at the South Pole, and with other techniques. The strongest signals will come from regions overabundant in WIMPs. These locations include the center of the Earth or the Sun, the Galactic Center, and dwarf satellite galaxies residing near or even inside the Milky Way. Experimentalists are hunting for signatures of dark matter annihilation from these areas and already have some intriguing results.

The WIMP annihilation process happens wherever the density of WIMPs is high enough for them to collide with one another. We have seen previously that WIMP annihilations in the early Universe are responsible for the WIMP miracle (Chapter 5). The number of relic WIMPs left in the Universe today can be computed and turns out to be exactly the required amount to account for dark matter. Today the average WIMP density is so low that most WIMPs in the Universe no longer collide with one another. However, in any location with an overabundance of WIMPs, the annihilation process will persist.

Locations containing the largest concentrations of dark matter would give rise to sizable annihilation rates and detectable signals in a variety of telescopes. In the 1980s I was involved in some of the first papers proposing such dark matter searches. The Milky Way Galaxy itself has an overdensity of WIMPs compared to the average for the Universe. We did computations of the annihilation rates for typical WIMPs in the Galaxy into photons, positrons, and neutrinos, and found that all of these processes could give interesting signals in detectors.

Two other places to look for dark matter annihilation products are the centers of the Sun and the Earth. A group of theorists pointed out that the Sun could capture some of the WIMPs that pass through it.[15] These would collect at the core of the Sun and annihilate to give a detectable signal. Following their work, I wrote a paper on indirect detection in Earth; the same process takes place in both objects.[16] One in 10 billion of the WIMPs traveling through Earth would hit a nucleus and lose enough energy to be captured. Then the WIMP is pulled to Earth's center by gravity. Eventually enough WIMPs will accumulate that they start annihilating with one another. Although the electron-positron pairs and the photons created by the annihilation cannot get out of Earth's core because of electromagnetic interactions with matter, the neutrinos from the annihilation can escape. They can make it to neutrino detectors on the surface of Earth that have already been built for other purposes. In my paper, I took advantage of the fact that the already-existing Irvine-Michigan-Brookhaven experiment hadn't seen any neutrinos coming out of the center of Earth. I was able to rule out Dirac neutrinos and scalar neutrinos (SUSY partners of ordinary neutrinos) as candidates for dark matter. Since that time, further searches for WIMP annihilation from the centers of Earth and the Sun have continued. As yet no signal has been found. Tighter bounds have been placed on WIMP models. The hunt for neutrinos coming to us from Earth's core and from the Sun is an ongoing effort.

Another direction in which to look for signatures of dark matter annihilation is toward the center of the Milky Way. Computer simulations of galaxy formation predict a high density of WIMPs in a small region near the Galactic Center. This WIMP abundance might be further enhanced by the 4-million-solar-mass black hole at the Galactic Center, which could pull in additional WIMPs and create a dark matter spike.[17] The spike would then produce even larger annihilation signals. However, the opposite effect could happen as well. If supermassive black holes form by mergers of smaller black holes, then these precursors might have knocked the dark matter particles away from the central regions of the Galaxy, much as a bowling ball would scatter ping pong balls (in the words of Santa Cruz physicist Joel Primack). In any case, the Galactic Center is expected to have excess dark matter. Thus dark matter hunters carefully study signals from the direction of the Galactic Center to look for signatures of WIMP annihilation. A more serious problem is the competition from the many astrophysical processes nearby, as they could produce unrelated energetic particles that would be seen in the detector. It is hard to disentangle a WIMP signal from an ordinary astrophysical one in the direction of the Galactic Center.

Dwarf satellite galaxies would produce cleaner signals. These small galaxies formed via the hierarchical structure formation process responsible for the existence of galaxies and clusters today. Inside our Milky Way Galaxy today there are still many smaller residual substructures, including dozens of dwarf galaxies that are 1,000 to 1 million times less massive than the Milky Way itself. For example, on the other side of the Galactic Center from us, the Sagittarius Galaxy is currently being gravitationally shredded apart and eaten by the Milky Way. Two tidal tails are being torn off in the process. The stars in one of the two tails have been observed by a number of groups to (possibly) pass not too far from the Sun. In a series of papers, my collaborators and I examined the consequences for dark matter detectors of the WIMPs accompanying the stars in one of these tidal tails.

Currently there are about 20 known dwarf galaxies orbiting inside our Galaxy, and new ones are still being discovered. As the galaxies are very faint, these observations are difficult. Two of the leading researchers in observational searches for ultrafaint dwarfs are Marla Geha of Yale University, who discovered the object called Segue, and Beth Willman of Haverford College, who has the dwarf Willman 1 named after her. The stellar population of Willman 1 consists mainly of old stars formed more than 10 billion years ago. Observations indicate that this dwarf galaxy is 1 ten-millionth as bright as the Milky Way and contains an enormous fraction of nonluminous mass, giving it a *mass-to-light ratio* of 800. The higher this ratio is, the larger will be the fraction of dark matter inside the object. These dwarf spheroidals have the largest fraction of dark matter inside them of all known galaxies. Consequently they are expected to have large rates of dark matter annihilation and are good places to look for signals in indirect detection experiments.

### Annihilation Products: Electron-Positron Pairs, Gamma Rays, and Neutrinos

Depending on the dark matter model, WIMP annihilations can produce a variety of different particles: quarks; leptons; W, Z, and Higgs bosons; or photons. Each of these annihilation products then decays or fragments into a chain of lighter and less-energetic particles. Typically the end-products of this chain are a combination of roughly equal numbers of neutrinos, electron-positron pairs, and high-energy photons known as gamma rays. These final particles from the annihilation each have energies typically about a tenth of the mass of the original WIMP. The goal then is to search for all three of these types of annihilation products. All of them are detectable, and there is a worldwide effort to search for these neutrinos, positrons, and gamma rays.

Given a particular type of WIMP, theorists can compute the numbers and energies of the particles that emerge from the annihilation. For example, if WIMPs are neutralinos from SUSY theories, then two WIMPs could annihilate directly to form pairs of bottom quarks, top quarks, tau leptons, or W bosons. These would further decay to lighter particles, producing a chain such as the one shown in Figure 7.7. For each decay, scientists can compute the predicted rates. Depending on the characteristics of the WIMP, detailed calculations can be made of the final photons, positrons, and neutrinos produced by these different annihilation channels. Observers can then search for these particles. If they are not found, then bounds are placed on the original WIMP

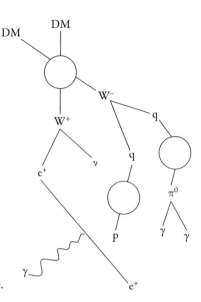

**FIGURE 7.7** Indirect detection looks for the final products of annihilation of Weakly Interacting Massive Particles (WIMPs; denoted by DM in the figure). Depending on the model, WIMP annihilations can produce Standard Model fermions, gauge bosons, or Higgs bosons, which then undergo a series of decays, for example, via pions ($\pi^0$), to a variety of final particles. Experiments in space and in the ice at the South Pole are searching for three types of final annihilation products: gamma-ray photons ($\gamma$), positrons ($e^+$), and neutrinos ($\nu$).

particles that started the chain. The hope is to detect a conclusive signal and discover the dark matter particle. Working backward from the predicted channels, scientists could further obtain detailed information on the nature of the WIMP.

The worldwide hunt for WIMP annihilation signals includes searches for all three WIMP annihilation end-products: positrons, neutrinos, and gamma-rays. Chapter 8 describes in detail the variety of experiments hunting for WIMP annihilation, including the PAMELA and FERMI satellites in space as well as the IceCube detector embedded in the ice at the South Pole. Chapter 8 highlights those experiments best poised to find signatures of dark matter. Particularly interesting are the unexplained signals of positrons and gamma rays that have been found in some of these detectors. Are these from WIMP annihilation? Do they herald the discovery of dark matter? The situation is interesting, to say the least.

# EIGHT

# Claims of Detection
*Are They Real?*

We live in exciting times in the hunt for dark matter. A host of experimental groups, using a variety of techniques, are reporting unexplained signals that may herald the discovery of dark matter. Yet other experiments see nothing. This perplexing situation is driving the competition among groups and engenders a sense of urgency to the search. This chapter explores the contradictory results and the attempts to reconcile them. The leaders of these experiments have strong personalities, and they are all competing to be the first to solve the dark matter problem and win a Nobel Prize.

The situation for dark matter detectors, both of the direct and indirect detection varieties, is fluid and exciting. The dark matter problem is not yet resolved, and every few months there is a new twist in the experimental situation. Even as I was writing this book, one new result after another came out that I rushed to describe before the book went into production. By the time this book is in print, there may be something new that I will wish I had been able to add to this chapter. The evolution is currently rapid, and a breakthrough may be expected in the coming years. My personal prediction is that we will have an answer to the dark matter problem very soon.

## Direct Detection in Underground Laboratories
### DAMA Annual Modulation

In laboratory WIMP experiments, including the Italian Dark Matter Experiment (DAMA), the signature of a dark matter particle is the interaction of a WIMP from the halo of the Galaxy hitting a nucleus in the detector. For more than a decade now, DAMA has been reporting claims of such a signal. Their work is based on an idea first proposed in a paper I wrote with Andrzej Drukier and David Spergel in 1986.[1] Thanks to this project, Andrzej calls us

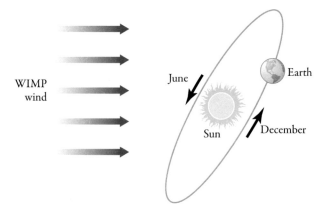

FIGURE 8.1 Earth's orbit around the Sun produces an annual modulation of the Weakly Interacting Massive Particle (WIMP) signal in detectors. The count rate is expected to have a peak in June and a minimum in December.

"The Three Musketeers." As described in Chapter 7, we suggested that experimentalists should search for an effect we called annual modulation. The count rates in WIMP direct detection experiments should vary with the time of year as a result of the motion of Earth around the Sun, whereas noisy background signals would not. As a consequence, finding this annual modulation in the data would be an important signature of dark matter detection.

The origin of the annual modulation of the WIMP signal is illustrated in Figure 8.1. Our Sun rotates in a roughly circular orbit around the center of the Milky Way with a speed of 250 kilometers per second, effectively moving into a wind of WIMPs. An analogy would be driving on a rainy day: as your car moves forward into the rain, you feel as though the raindrops are headed directly into the windshield. The flattened disk of the Galaxy, containing the Sun, rotates around the center, while the giant (roughly) spherical halo containing the WIMPs is essentially nonrotating.[2] Although typical WIMPs travel just as rapidly as the Sun, their motions are in arbitrary directions in the rest frame of the Galaxy, that is, from the perspective of a stationary observer situated at the Galactic Center. On average we are moving into a steady wind of WIMPs.

Then, as Earth circles around the Sun on its yearly orbit, our motion relative to the wind changes with an annual cycle. Earth revolves around the Sun at 30 kilometers per second (67,000 miles per hour). The plane of Earth's orbit is at a 60 degree angle from the plane of the Sun's motion in the Galaxy. Drukier, Spergel, and I computed that the WIMP count rate should exhibit a sinusoidal variation over time: the dark matter signal should peak in early June and reach its lowest point in December. The magnitude of the variation should be roughly 5% of the total count rate. In a follow-up paper I wrote with collaborators, we showed that experimentalists should be able to extract evidence for such a modulation from their data even in cases where the background noise is larger than the WIMP signal itself.[3]

In the 1990s, the DAMA group headed by Rita Bernabei at the University of Rome "La Sapienza" designed an experiment to look for WIMP annual modulation with the goal of discovering dark matter. Rita is a smart, excitable, charismatic personality with a top scientific crew. The DAMA experiment is situated in the Gran Sasso Tunnel underneath the Apennine Mountains in Italy. The detectors are made of ultrapure sodium iodide (NaI) crystals grown specifically for DAMA. A major requirement is that the crystals have very low radioactive contamination. Neutrons from radioactive decay would produce the same signal in the detector as dark matter particles, and it is impossible to tell them apart.

The detector registers a signal when an incoming particle hits one of the sodium or iodine nuclei in the crystals. The energy deposited by a WIMP interaction causes the crystals to *scintillate*; that is, they absorb the particle's energy and reemit flashes of light. Sensors called photomultiplier tubes pick up the scintillation light and convert it to electrons. Each electron is then attracted to an electron multiplier, which creates more and more electrons. The resultant cascade of electrons provides a measurable current. Photomultiplier tubes can be designed to be so sensitive that they can even detect single photons.

The conversion of light to electrons that takes place in the photomultiplier tubes (phototubes) is known as the *photoelectric effect*. Discovered in the late nineteenth century, this effect played an important role in the development of quantum mechanics. It was the first proof that light could be thought of as particles called photons. In 1905 Albert Einstein mathematically described the photoelectric effect as the absorption of photons and the reemission of photoelectrons. The new concept of wave-particle duality, the notion that all of matter can be depicted equally well as particles or as waves, became central to quantum mechanics. In 1921 Einstein won the Nobel Prize for this work.[4] Photomultiplier tubes using this effect are at the heart of DAMA and many other experiments today.

In 1997, at the COSMO conference at Ambleside in England, Pierluigi Belli, Director of the National Institute of Nuclear Physics in Rome, presented the first preliminary results from the DAMA experiment. At that time the group had very little data but was starting to see a difference between the count rates in summer versus winter. The response from the cosmology community was skeptical, primarily because there simply weren't enough data: the group had only 10 days of data in the summer and a month's worth in winter. Neil Spooner of the British Zoned Proportional Scintillation in Liquid Noble Gases (ZEPLIN) experiment gave a short talk at the conference in which he listed 10 alternate explanations of the DAMA data that had nothing to do with dark matter. The possibilities included seasonal temperature variations in the tunnel and radioactive decay of radon. The danger of presenting tantalizing hints when the statistics are still too low to be persuasive is that it engenders suspicion in the scientific community. The doubts about the results can persist even after more data are found years later. Yet the temptation to validate the hard work of many years with at least some interesting results is understandable, especially when future funding may depend on it.

The COSMO conference where Belli announced the first DAMA results took place in the 100-year-old stone buildings of St. Martin's College, roughly a 5-hour drive north of London. The participants stayed in dorm rooms that were vacant during student holidays. One of the buildings had a small quaint bar, which the bartender kept open late for us, so that we could enjoy the local Scotch and play darts. I, for one, got frustrated and found myself hurling the darts like baseballs (my favorite sport in childhood). The conference was also notable for an organized afternoon of Scotch tasting and the accompanying songs led by Nancy Abrams, "We all live in an expanding universe, expanding universe, expanding universe" (sung to the tune of the Beatles' "Yellow Submarine"). The town of Ambleside is in the Lake District, near Scotland, and although the lakes have inspired beautiful romantic poetry, they are freezing. I dared an Austrian participant to jump into the lake at midnight from a boat dock. After he dove in I followed, but we both clambered back out of the icy water immediately. We had borrowed a car to get to the dock, and on the way back I inadvertently drove on the wrong (right) side of the street and wondered why the "idiots" coming at me were about to hit me.

After the initial announcement in Ambleside in 1997, the DAMA experiment continued to run and has accumulated more than 13 years' worth of data. The evidence for annual modulation is undeniable. The statistical significance is at the level of a 9-sigma detection, which means that the likelihood of it's being just a lucky coincidence is down at less than 1 in 1 billion. In addition, the 10 items listed by Spooner as alternate explanations for the modulation have all been checked. The members of the DAMA group are excellent experimentalists, though not the best presenters at conferences, and they have systematically eliminated all the possible spurious signals suggested by Spooner.

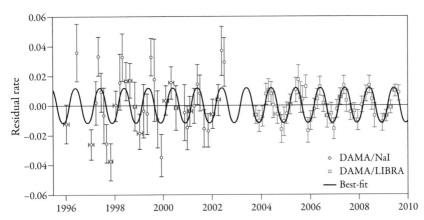

**FIGURE 8.2** Direct detection count rates over more than a decade from the DAMA experiment in Italy, showing data points with error bars. The time dependence of the signal is exactly consistent with the annual modulation expected for a Weakly Interacting Massive Particle signal stemming from Earth's motion around the Sun (theoretical predictions are shown by the solid line). The average count rate over the whole time period has been subtracted off, so that only the residuals are plotted. As predicted, the count rate peaks in June and has a minimum in December. *From Bernabei, R., P. Belli, F. Cappella, R. Cerulli, F. Montecchia, F. Nozzoli, A. Incicchitti, and D. Prosperi, et al. 2003. "Dark Matter Search." Rivista del Nuovo Cimento 26: 1; Bernabei, R., et al. [DAMA and LIBRA Collaborations]. 2010. "New Results from DAMA/LIBRA." European Physics Journal C 67: 39.*

If it really is seeing WIMPs, then by now DAMA has observed nearly 1 million WIMP scattering events in the detector. The data are consistent with an interpretation in terms of annually modulated WIMPs. The count rate has the right phase, with a maximum signal in June and a minimum in December. The group has now published the *spectrum* of the results (that is, the number of events at different energies), and the shape of the spectrum is consistent with a WIMP signal. In 2003 the DAMA results passed another critical test. The detector was upgraded to a new design (DAMA/LIBRA) with more detector material (250 kilograms of NaI crystals). Using the improved technology, the group repeated the experiment, and the data continued to show the same pattern of annual modulation (Figure 8.2).

For more than a decade, DAMA has reported exactly the annual modulation we predicted. Unfortunately, however, the situation is still confusing, in part for sociological reasons. The experimentalists are unwilling to share their data with outsiders, a decision that makes the scientific community skeptical of the conclusions. The tradition in particle and astrophysics experiments is to make the data public soon after publishing the first results. Then any scientist, even outside the group, can examine the data to look for inconsistencies or alternate interpretations. As experimentalist Frank Avignone said to DAMA's leader (when she declined at a conference to release the data on the web), "But Rita, we are all your friends!"[5] On the DAMA webpage is the following quote from Rudyard Kipling:[6] "If you can bear to hear the truth you've spoken twisted by knaves to make a trap for fools, you'll be a Man my son!"

The other reason for doubt is that the DAMA data appear to conflict with results from competing experiments. In the time that DAMA has been accumulating data, other groups have also been making measurements, as described in the rest of this chapter. Three of these, CoGeNT, CRESST, and CDMS, show preliminary evidence of a positive signal that may agree with DAMA, but the rest report no signs of a dark matter signal. This lack of events is known as a *null result* and places limits on the possible types of WIMPs. The strongest bounds include those from the XENON experiment in the Gran Sasso Tunnel and the LUX experiment in South Dakota. Yet it is dangerous to directly compare the results, because these detectors are made of profoundly different materials: XENON of liquid xenon and DAMA of NaI crystals. Thus it is possible that the cause of the discrepancy among the results is the difference in the detector materials—much like comparing apples to oranges.

Chris Savage, Paolo Gondolo, Graciela Gelmini, and I tried to make sense of this puzzling situation. Why does DAMA see signals compatible with dark matter, while so many other experiments do not? The standard way of reporting the results of these experiments is in terms of the *cross-section* that illustrates the scattering strength implied by the data. This quantity is a measure of the likelihood that an incoming WIMP will strike a nucleus. One way to visualize a cross-section is to draw a circle around the target nucleus, perpendicular to the line between the nucleus and the WIMP. Any dark matter that comes inside that circle has effectively hit the nucleus, whereas any WIMP outside it just passes on by. A bigger circle—a larger cross-section—implies that an incoming WIMP has a better chance of hitting the nucleus. A larger cross-section leads to more frequent interactions.

In interpreting their data, scientists differentiate between a variety of possible types of weak interactions between WIMPs and nuclei. As described in Chapter 7, spin-independent interactions scale as nuclear mass squared, whereas spin-dependent interactions rely on the nuclear spin. The different types of scattering should produce different count rates in the detectors.

Figure 8.3 shows a plot of cross-section versus WIMP mass for the case of spin-independent interactions. To obtain the highlighted regions in the plot, the count rates have been divided by the square of the atomic mass of each

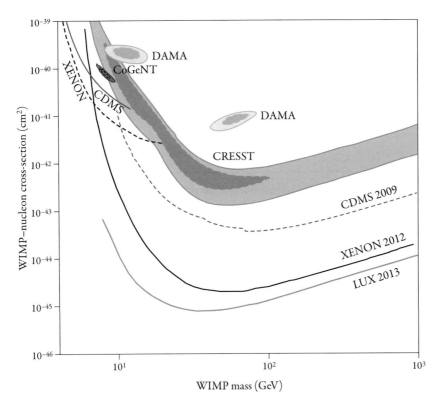

**FIGURE 8.3** Comparison of results from a variety of laboratory dark matter experiments for the case of spin-independent Weakly Interacting Massive Particles (WIMPs). The solid zones show regions compatible with the signals observed in the DAMA, CoGeNT, and CRESST experiments. The curves show limits placed by null results (zero signal) of the CDMS and XENON experiments; only regions below these curves are allowed. The null results are clearly in tension with the positive signals, implying either experimental error, incorrect astrophysical halo models, or wrong particle physics assumptions. *Plot made by Joachim Kopp for inclusion in Freese, K., M. Lisanti, and C. Savage. "Colloquium: Annual Modulation of Dark Matter."* Reviews of Modern Physics *85(4): 1561.*

detector to eliminate this factor from the comparison. The advantage of presenting the data in this way is that the results of many different experiments can be simultaneously plotted and then compared.

Figure 8.3 plots the experimental situation as it stood at the beginning of 2014. Every few months it needs to be updated with the latest results. One can see the regions consistent with DAMA data as well as the results from other experiments for the case of spin-independent WIMP interactions with nuclei. Either of the two regions labeled "DAMA" in the figure would be compatible with DAMA data. The reason that there are two possible regions is that the experiment contains two different detector elements, sodium and iodine. Iodine, with atomic mass A = 127, is much heavier and harder to budge than sodium, with A = 23. When reconstructing the interaction from the data, the experimentalists found two alternatives. The observed DAMA data could be due to WIMPs with a mass of roughly 80 GeV (80 proton masses) scattering off of iodine nuclei, or they could be due to WIMPs of roughly 10 GeV (10 proton masses) scattering off of sodium nuclei. These two possibilities are shown as the roughly circular regions marked "DAMA" in the plot.

In addition to the two regions compatible with DAMA data, Figure 8.3 also shows the upper limits resulting from other experiments. To be compatible with the lack of signal (null results) in the CDMS, XENON, and LUX detectors, WIMPs would have to have masses and cross-sections that would put them below the XENON and CDMS curves in the plot. Researchers ponder the perplexing riddle of why DAMA sees positive signals while other experiments do not. The consensus is that the heavier of the two possible DAMA WIMP mass regions—the 80-GeV WIMPs—is in drastic conflict with the null results of all other experiments and is probably excluded as a possibility. In 2009 (prior to the LUX results described below), my collaborators and I did a careful statistical analysis and concluded that for some types of particle interactions, low-mass spin-dependent candidates—roughly 10-GeV WIMPs—could be simultaneously consistent with all the data and survive as a possible dark matter detection.[7] Other groups came to the same conclusion. The experiments with the strongest negative results at 10 GeV were not designed for this mass range, and their limits may be incorrect. If the DAMA modulation really is due to WIMPs, then the dark matter particle weighs roughly 10 times as much as a proton.

Other physicists argue that the DAMA data have nothing to do with WIMPs. Alternative explanations include variation of the ambient temperature in the Gran Sasso Tunnel housing the experiment or fluctuations of decaying radioactive material in groundwater. Another possibility is contamination by muons, particles created in Earth's atmosphere that could make it into the detector. The muon rate in the tunnel is known to annually modulate because of temperature variations in the upper atmosphere. The DAMA experimentalists have studied these issues and claim that their measurements rule them out as an explanation, but not everyone is convinced.

As more data came in and the results continued to hold up, the possibility that DAMA has made a discovery was taken more and more seriously. Theorists who initially scoffed at the experiment feverishly tried to create WIMP models consistent with both DAMA and the null results from the other experiments. Even some experts from other experimental groups became cautiously optimistic. Super-careful Frank Calaprice at Princeton University surprised me in June 2013 when he told me that he was interested enough in the DAMA data to build a new experiment together with a postdoctoral fellow to check the results.[8] However, he warned that there is one explanation compatible with all the existing data: all the positive signals in experiments, including DAMA, could be caused by background noise rather than by dark matter particles.

### CoGeNT in the Soudan Mine

Starting around 2010, the interest in light WIMPs—roughly 10 times as massive as protons—heated up tremendously. Not only could particles of this mass explain the DAMA data, but several other experiments also reported tentative evidence of similar results.

Juan Collar leads the Coherent Germanium Neutrino Technology (CoGeNT) collaboration at the University of Chicago. CoGeNT is a small experiment, consisting of 440 grams of germanium detectors that look like hockey pucks. Although it has less total mass than other laboratory dark matter experiments, CoGeNT is particularly well suited to look for low-mass WIMPs weighing 10 GeV. When they strike detectors, these particles deposit small amounts of energy. Most of the other experiments are simply not sensitive to these low energies. In contrast, CoGeNT is designed to look for them. In 2010 the collaboration reported unexplained events that might be attributed to WIMPs. The initial analysis indicated that these results could be consistent with the 10 GeV WIMPs that DAMA appeared to be seeing. The tricky part is that CoGeNT is unable to differentiate nuclear recoils (which could be from WIMPs) from electron recoils (which would be considered background noise). Some of the contamination is well understood: known radioactive elements give rise to peaks in the data, including nickel-56, zinc-65, and vanadium-49 (the numbers after the elements indicate the material's atomic mass: total numbers of neutrons plus protons). The experimentalists can subtract these peaks, leaving behind signals that could be from WIMP interactions.

Then the situation became even more exciting. In May 2011 I encountered Juan in the lobby of the Hyatt Hotel in Anaheim where the American Physical Society meeting was being held. He was en route to taking his kids to Disneyland. As he gave me a big hug, he told me that the next day he would be presenting a talk with results on annual modulation in the CoGeNT data. In the words of another friend, "Juan is the straw that stirs the drink." I begged him to tell me the statistical significance of what he had found, but he declined to reveal his results before his lecture. I got overexcited. From a third person,

I heard a misleading rumor that the significance was higher than it was. I concluded (wrongly) that CoGeNT had confirmed the DAMA data, and that WIMPs had been discovered. I contacted several reporters as well as wrote a blog for the World Science Festival in New York in which I would participate in June.

The reason I got so excited was that the true test of a discovery will be complementary detection in multiple experiments. DAMA is in Italy, whereas CoGeNT is in the Soudan Mine in Minnesota. DAMA is made of NaI; CoGeNT uses germanium. Confirmed signals in both the DAMA and CoGeNT experiments would provide conclusive evidence of WIMPs.

Juan is a fiery person with a sharp wit. Regarding his chief competitors, he says, "I'm a Spaniard caught between two Italian women!" He is referring to Rita Bernabei, who leads the DAMA experiment, and Elena Aprile, the group leader of XENON (Figure 8.4). That night before his presentation at the American Physical Society meeting, I got an email from him, saying "I gotta go back to my punching ball, getting ready for showdown tomorrow." He spent half of his talk in Anaheim criticizing the XENON experiment, which has null results apparently incompatible with his data. Nevertheless, Laura Baudis gave an excellent presentation of the XENON results, and over the next few years the XENON experiment has among the best chances of detecting WIMPs.

It turned out that the statistical importance of the CoGeNT annual modulation was lower than I had thought. The amount of data wasn't enough to say anything conclusive. Because of a fire in the Soudan Mine in Minnesota where the detector is located, CoGeNT had been forced to shut down. As a consequence, the modulation data Juan presented at the American Physical Society meeting were only marginally statistically significant.

After the conference Juan wrote me, "Probably next week I'll know if the detector is still usable. If that is the case, I may have to stop being an agnostic in just a few short months. Cheers and stay tuned!" Luckily, the experiment was not destroyed and has since continued taking data.

In the meantime, several other experiments came out with analyses that disagreed with the CoGeNT results. The CDMS and XENON experiments both reexamined their data down to the low energies of CoGeNT and claimed to rule out its results. These are two of the major experimental efforts of the dark matter community and will be described further below. The CoGeNT group then reanalyzed its existing data as well as obtaining new data, and continued to argue that their signals are correct. The arguments between Juan and various representatives of the groups have become somewhat vitriolic, but in the end, in science there is only one right answer. The questions remain contro-

"I'm a Spaniard caught between two Italian women."

FIGURE 8.4 (A color version of this figure is included in the insert following page 82.) "I'm a Spaniard caught between two Italian women." Juan Collar of CoGeNT (center), Rita Bernabei from DAMA (left), and Elena Aprile from XENON (right) are leaders of three of the principal dark matter experiments. *(Left) Rita Bernabei, ONFS; (middle) Juan Collar and the University of Chicago; (right) Richard Perry/*New York Times.

versial: is the interpretation of the CoGeNT experiment in terms of WIMPs correct? Is CoGeNT consistent with DAMA?

Regarding the future of dark matter experiments, Juan says, "It's not CDMS that keeps me up at night, it's the Chinese." Here he is referring to the upcoming PandaX experiment, which will be located in China in the deepest laboratory on Earth. Because of the depth of its location, this experiment will have the lowest contamination from neutron background, the worst offender in producing spurious signals in detectors. The Chinese government plans to dedicate substantial financial support to these experiments just as U.S. science budgets are shrinking drastically. I do agree that it would be bad for U.S. science if the techniques that were painstakingly developed over the past 20 years in the United States were scaled up in another country and the Nobel Prize were to go elsewhere.[9]

## The CRESST Experiment

In spring of 2012 the 10-GeV WIMP story heated up even further. A third dark matter experiment, the Cryogenic Rare Event Search with Superconducting Thermometers (CRESST), also reported unexplained events compatible with a WIMP mass of roughly 10 GeV.[10] The new results initially appeared to be consistent with the hints of low-mass WIMPs in DAMA and CoGeNT. Developed at the Max Planck Institute in Munich, CRESST was deployed in the Gran Sasso Tunnel in Italy. This is the same underground laboratory that houses the DAMA and XENON experiments. The CRESST detector is made of $CaWO_4$ scintillator, consisting of calcium, tungsten, and oxygen. The advantage of using three different detector elements is that they can be used to give complementary information about the WIMP particle. In the long run, comparison of results in different energy ranges should allow for powerful checks of a WIMP signal.

CRESST reported an excess of low-energy events of unknown origin with a statistical significance of 4 sigma, corresponding to 99.994% probability that these events are not arising purely randomly. Yet this experiment suffers from radioactive contamination originating from the clamps holding the detectors in place. The radioactivity could mimic a WIMP signal. Computer simulations estimate the expected number of false signals, but the required extrapolation from measured values is difficult. A redesign of the detector has eliminated this background source. CRESST's Leo Stodolsky writes: "There are two types of new detectors installed. And so far they are totally background free—no events in the regions of interest. We'll see how this continues but this is a big step forward."

Future CRESST runs should clarify the origin of the current excess. In a previous incarnation, the CRESST detector had a different source of contamination. The sapphire in the detector buckled under pressure, with the consequence that the group ended up publishing papers on the cracking properties of sapphire rather than on dark matter. These experiments are very difficult!

## The CDMS Experiment

The pioneering experiment that started low-temperature dark matter searches was the Cold Dark Matter Search (CDMS) experiment. Beginning in the late 1980s, CDMS has remained at the forefront of the hunt for dark matter. This group was led by Bernard Sadoulet (Figure 8.5), David Caldwell, and Blas Cabrera at its inception and has trained some of the best astrophysicists in the United States today. The headquarters of the group is in the San Francisco Bay area, and institutions all over the United States have become participants. It is physically located directly next to CoGeNT and is made of two materials— germanium (the same as CoGeNT) and silicon.

This group is famous for being extremely careful. The experiment has minimal background noise by design. The detector measures both the heat and the charge deposited in the detector by any incoming particle. Unlike DAMA or CoGeNT, but similar to CRESST, the CDMS experiment can differentiate events caused by electrons from the nuclear scattering events of interest.

FIGURE 8.5 A pioneer of dark matter searches: Bernard Sadoulet of the CDMS experiment. *Dennis Seitz.*

Because electrons are charged, they ionize the detector much more than WIMPs would.

CDMS started taking data in the late 1990s with a detector prototype situated on a tabletop. The CDMS team showed that their technology was already able to start setting limits on WIMPs. Then CDMS-II moved into the Soudan Mine in Minnesota, one of the deepest laboratory sites on Earth. The sensitivity of the experiments has improved by about a factor of 1,000. It is physically located directly next to CoGeNT and is made of two materials, germanium (the same as CoGeNT) and silicon. CDMS-II still found null results for a WIMP signal, in apparent conflict with claimed detections in other experiments.

As shown in Figure 8.3, the null results from both CDMS and XENON (discussed below) naively appear to rule out the positive results from all three experiments, DAMA, CoGeNT, and CRESST. The question is how to interpret the disagreement among the different groups. Because CDMS is made of germanium and silicon whereas DAMA is made of NaI, comparison of these two experiments is tricky, and it is possible that they are both correct. In contrast, both CDMS and CoGeNT use a germanium as target material and are situated next to each other in the same mine. Because they have the same target material, their measured rates can be directly compared. Motivated by the CoGeNT results, the CDMS team reanalyzed its germanium data down to the low energies appropriate for 10-GeV WIMPs. This study placed strong constraints on the low-mass WIMP region. For the standard spin-independent scattering (where the signal scales with the square of the atomic mass, as discussed in Chapter 7), the CDMS experimentalists argued that their bounds disfavor a WIMP explanation for DAMA/LIBRA and CoGeNT. Possible compatibility would have required at most half the events from CoGeNT to be from WIMPs. Then the CoGeNT group reanalyzed the events recorded near the surface of their detector and found that many of the signals they first took to be WIMPs were in fact due to radioactivity. They modified their signal down to a level where it became compatible with CDMS bounds.

The CDMS experimentalists then examined their germanium data set to look for annual modulation in the rate. The modulation amplitude measured by CoGeNT should have produced a whopping modulation in the same energy bins in CDMS, ranging from almost no counts in December to a large number in June. But CDMS found no evidence for modulation. They claimed that this result again disfavored a low-mass WIMP explanation for the CoGeNT modulation.

In mid-April 2013, we hosted a conference at the University of Michigan titled "Light Dark Matter" on the subject of 10-GeV WIMPs (the lead organizer was Kathryn Zurek). Enectalí Figueroa-Feliciano from MIT, a member of the CDMS collaboration, was the second speaker at the meeting. The night before his talk, he emailed me and said I wouldn't want to miss it. Though I had been recovering from a concussion (due to a head-on collision with another swimmer), I arrived in time to hear his results.

What a surprise! In his talk he announced that CDMS also had three events compatible with 10-GeV WIMPs. The CDMS experiment is made not only of germanium but also contains some silicon. In two of their eleven silicon detectors, the group found nuclear recoils that were clearly differentiated from any noisy electron signals. As conference organizers, we couldn't ask for more than having a major announcement at our meeting. The speaker later told me that we had motivated him to speed up the data analysis. He couldn't imagine going to a conference exactly on the topic of his findings and then have the results come out a few weeks later. Andrew Grant from *Science News* reported on these events.[11] He quoted me as saying, "I'm more excited than I should be, but I can't help it." I think that statement pretty much sums up the sentiment of all the attendees at the meeting. A small number of events right at the very

FIGURE 8.6 Dark matter women at the University of California, Los Angeles, Dark Matter Meeting, February 2012: Laura Baudis (XENON), the author, Elena Aprile (XENON), and Lisa Randall (theorist from Harvard). Behind Lisa is Rick Gaitskell (LUX).

lower limit of the detector's sensitivity were not enough to prove anything, but the consistency among all the different experiments looking for WIMPs in the 10-GeV mass range is what made the new CDMS data interesting.

### A Story of Early CDMS Results

Every 2 years, David Cline of the University of California, Los Angeles, runs the UCLA Dark Matter Meeting at the Marriott Hotel in Marina del Rey, about half an hour from the Los Angeles airport.[12] It's my favorite conference, period, and many of us look forward to it every time (Figure 8.6). It's an extravaganza. The hotel, located about a mile from Venice Beach, overlooks the Pacific Ocean. The talks are given either in the downstairs ballroom or on the ninth floor with three of four walls made of floor-to-ceiling windows with a view of the ocean. The talks start at 9 in the morning and are scheduled until midnight! Dave Cline gets requests for far too many talks for a 3-day period, and he overschedules, plain and simple. Yet every single one of the talks is worth attending. Although the participants complain, in the end it is always an extremely successful meeting. If you want to eat, to some extent you have to pick and choose which talks you will skip. The conference also serves breakfast, lunch, dinner, and endless coffee and cookies—it's hard not to be eating constantly.

As a break from talks, some of the attendees rent bicycles right across the street from the hotel and ride the bike path over to the scene in Venice. The sidewalk is full of joggers, rollerbladers, and people on bicycles—many in outrageous outfits. Teenagers on a skateboard ramp do tricks and fall onto the concrete without seeming to get hurt. Jugglers and other street performers collect large crowds. My favorite is Muscle Beach, an outdoor gym right on the beach, where Arnold Schwarzenegger apparently used to work out. Every time I attend the conference, I buy a Muscle Beach t-shirt and at least two pairs of purple sunglasses. About a 5-minute walk from the conference hotel is the Baja Cantina, a Mexican restaurant that for many of us is the main attraction in the evenings. We don't go there for the food; we go there for the margaritas. We order the extra-large size with six straws per glass, and we all get pretty excited.

One of the funniest scenes was in the year 2000. We all knew that CDMS was planning to release results with the potential of confirming or ruling out the DAMA annual modulation. The CDMS group members were very secretive so as not to let out their results ahead of the scheduled press release at the conference. Rick Gaitskell, now a professor at Brown University, but at the time a graduate student looking for a postdoctoral position, had been selected to give the talk. Originally from London, he comes from a British political family. He is extremely passionate about science, and along with the other experimentalists, hopes that a professional life dedicated to these searches proves to give rise to major discoveries. Currently he is a leader of the Large Underground Xenon (LUX) experiment.

The night before his talk, members of the CDMS experiment were clustered together in the large ballroom, while Rick practiced his presentation in secret. A group of us decided to heckle him. After a large number of margaritas at the Baja Cantina, we found a bottle of sherry, snuck into the back of the ballroom, and laughed so that he would know we were there. Rick took one look at us and shouted across the room in proper British: "Oh bugger off. You're going to spoil the fun!" We left and went to bed.

The next day Rick gave his talk to an audience of 300 experts and showed that CDMS had found no evidence of WIMPs. He argued that their results disproved DAMA. The room erupted with questions and disagreement. At that point the CDMS experimentalists were still running the experiment above ground, on a tabletop in Stanford. So the big issue was whether they understood their neutron backgrounds well enough to make any statement at all. They claimed that computer simulations extrapolating from their multi-neutron events enabled them to get a handle on this issue. I was interviewed by James Glanz of the *New York Times* about my opinion on this. Since I'm not an experimentalist, I told him I'd just have to sit back and wait until it all shakes out. He wrote an article on this controversy between DAMA and CDMS.[13] He likened the dark matter searches to *Waiting for Godot,* the absurdist play by Samuel Beckett in which two men endlessly await the arrival of Godot, but Godot never appears. Since the time of that article, the dark matter field has seen a major resurgence of excitement thanks to hints of detection in so many different experiments.

## Bubble Chambers

A completely different approach to direct detection of dark matter uses bubble chambers, vats filled with *superheated* fluid—a liquid heated to just above its boiling point, at a temperature where it should be boiling, but is prevented from doing so. The fluid is hot enough to convert to a gas, yet it is still kept in its liquid phase. Even the tiniest amount of energy added to the fluid will cause easily visible bubbles to form. WIMPs striking the detector would have exactly that effect: they would deposit enough energy to induce the nucleation of a series of bubbles observed by the experimentalists. Juan Collar, the head of CoGeNT, also leads COUPP (the Chicagoland Observatory for Underground Particle Physics). This bubble chamber is filled with superheated $CF_3I$, an ingredient found in fire extinguishers. Two other groups are using the same technique: SIMPLE (Superheated Instrument for Massive ParticLe Experiments) and PICASSO (Project in Canada to Search for Supersymmetric Objects). These experiments have searched for dark matter and reported null results. There is no WIMP signal in the bubble chambers. Their results are in serious tension with interpretation of DAMA data in terms of spin-dependent WIMPs, whose interactions with matter depend on the spin of the nucleus (see Chapter 7). Bubble chambers will continue to search for signals of WIMPs. The upcoming collaboration PICO will have fantastic sensitivity to light WIMPs.

## Liquid Xenon and Argon

Elena Aprile of Columbia University is the leader of the XENON direct detection experiment. Originally from Italy, the stylish Aprile is an ambitious group leader. Involved for many years in the development of low-temperature noble liquid detectors, she realized she could turn her expertise to the dark matter problem. The *noble* elements are those situated in the farthest right column of the periodic table. They are chemically very stable and do not easily react with any other elements. In their gaseous form, they are often used for storefront lamps. Three noble elements, xenon, neon, and argon, are being explored as possible WIMP detectors. In the liquid phase, all three would be sufficiently dense to make good dark matter targets. These noble liquids would scintillate (produce light flashes) as WIMPs moved through them.

Another key player in the XENON collaboration is Laura Baudis (see Figure 8.6). Since 2007 Laura has led the particle astrophysics group at the University of Zurich in Switzerland.

The XENON detector currently consists of 100 kilograms of liquid xenon as the target material for the WIMPs. XENON is a two-phase experiment. Liquid xenon both scintillates and becomes ionized when hit by particles. The impact of a dark matter particle produces two flashes of light detectable in photomultiplier tubes (similar to those used by the DAMA experiment). The first flash is due to photons created directly in the collision. The second is caused by electrons liberated in the collision that drift in an electric field toward a layer of xenon gas. The ratio of these two signals depends on the nature of the incoming particle and can differentiate between a WIMP and an uninteresting background particle.

The XENON experiment has seen no definitive evidence for dark matter and has instead placed among the most stringent limits on WIMPs. Of the two mass regimes compatible with the DAMA data, XENON has ruled out the higher-mass 80-GeV WIMPs. Motivated by recent interest in the lighter WIMP masses, XENON reexamined its data at lower energies. As mentioned above, its most recent results are in serious tension with the 10-GeV WIMPs preferred by DAMA, CoGeNT, and CRESST. Yet the comparison of these experiments remains controversial. First there is the issue that the detectors are made of a variety of different target materials. Another major factor is that some researchers question the validity of XENON's results in the low-energy range relevant to 10-GeV WIMPs. XENON was designed to study heavier particles, and its sensitivity at the low-mass end is under debate. As Aprile's vats of xenon are upgraded to 1 metric ton (1,000 kilograms) of detector material, the dark matter community eagerly awaits the next results of the XENON experiment.

Liquid xenon detectors are a new entry to the dark matter wars. They are large and easily scalable to become even larger. But precisely because they are so massive, they are prone to radioactive background contamination due to neutrons. Other groups also plan to use liquefied noble gases in underground facilities as dark matter detectors. These collaborations include PandaX, DARKSIDE, and DEAP/CLEAN. PandaX will be located in the Jin-Ping II Hydropower tunnel in China, the deepest laboratory site on Earth. It is in the middle of an 18-kilometer-long tunnel under 2,400 meters of rock. PandaX plans to build a metric ton–scale liquid xenon detector. Other experiments use argon or neon as the target material. Frank Calaprice, who was my physics professor during my sophomore year at Princeton (and who looks like Gregory Peck), is one of the leaders of the DARKSIDE experiment using liquid argon detectors. The DEAP/CLEAN group hopes to scale an argon and neon detector up to 50 metric tons in mass. Proposed 10-metric-ton xenon experiments include LZ (derived from a combination of LUX and ZEPLIN) and DARWIN. The future of the field of direct detection will see the development of these enormous next-generation dark matter experiments.

In fall of 2013, the Large Underground Xenon Experiment (LUX), which also uses liquid xenon as its target material, made a major announcement. LUX is led by Rick Gaitskell and Tom Shutt. This detector has been taking data in the Sanford Underground Research Facility, an underground laboratory 1 mile under the earth, in the Black Hills of South Dakota. In October 2013, LUX reported results of their first WIMP search dataset, taken from April to August 2013, and presented an analysis of 85.3 days of data with 118 kilograms of liquid xenon. They reported null results that placed limits on low-mass WIMPs that were 25 times more stringent than those from the XENON experiment. Again the question is whether their results can be trusted at such low energies so near the threshold of their detector (below the energy threshold, the detector is not sensitive enough to pick up any signals). Their analysis was nevertheless very solid. The LUX results are shown in Figure 8.3. WIMPs would have to have masses and cross-sections that would put them below the XENON and LUX curves. These results are the latest surprise in the dark matter story—the situation will remain fluid until the dark matter problem is resolved!

Does LUX rule out DAMA, CoGeNT, and CRESST results? Some believe so. If correct, LUX results would eliminate the low-mass 10-GeV WIMP scenarios that became so popular over the preceding 3 years; the WIMP scattering strength (cross-section) would have to be lower than the value consistent with the positive signals from the three other experiments. If LUX is right, the constraints are so strong that even nonstandard WIMP models cannot seem to evade them. WIMPs could still have any mass, ranging from 1 GeV to 10 TeV, but would need to have a lower cross-section for scattering with nuclei.

In contrast, other physicists are not yet convinced by the LUX results. A *Scientific American* article, written on the same day as the release of the LUX results, states that "The other teams, however, are not ready to concede defeat. . . . Blas Cabrera of Stanford University, who leads the CDMS project, also maintains that what his project has seen may still prove to be dark matter. 'It is unlikely that LUX has ruled out the entire region of interest' for low-mass WIMPs because xenon is not as sensitive as other materials to dark matter in that mass range, he says."[14]

These are dark matter wars! Future experiments will tell.

## Light WIMPs

To me it is funny that the 10-GeV WIMP is back. The original calculations I made with my collaborators in the 1980s were specifically for such light WIMPs. Yet almost immediately after our work, particle accelerator results ruled out the simplest SUSY dark matter candidates with masses less than about 30 GeV.[15] Theorists and experimentalists alike readjusted their expectations for the ideal WIMP mass and started searching for heavier particles, in the range of 100 to 1,000 GeV. Most experiments have been geared to this heavier range and are not sensitive to the lighter masses. Starting with DAMA, however, experimental results resurrected light dark matter. The 10-GeV WIMP that we considered in the first place became the new favorite as DAMA, CoGeNT, CRESST, and CDMS all saw hints of such WIMPs. Yet the liquid xenon experiments find no signal. New experimental designs may be needed to test the light dark matter hypothesis. The next round of experiments will determine whether these low-mass WIMPs are the solution to the dark matter problem. Of course, another possibility is that all the interest in light dark matter will vanish, and the community will focus instead on the possibility of heavier WIMPs, up to 10 TeV in mass.

## Indirect Detection of WIMP Annihilation: Positrons, Neutrinos, and Gamma Rays

Indirect messengers of dark matter—photons, neutrinos, and electron-positron pairs from WIMP annihilation—provide another avenue to dark matter discovery. Because many WIMPs are their own antiparticles, they annihilate among themselves wherever they occur in large numbers; as a result of the annihilation, they produce these messenger particles. As described in detail below, in our own Galaxy, the regions of large WIMP density are the Galactic Center, the small dwarf spheroidal galaxies residing inside the Milky Way, and the entire Galaxy as a whole. These locations contain enough WIMPs to produce detectable signals of the three types of dark matter annihilation products. A variety of experiments are taking data to search for all three possible messenger signals. Unexpected results, possibly indicative of dark matter, have been found in indirect detection experiments and have generated much speculation and excitement.

The past few years have seen a flurry of excitement due to an unexpectedly large positron flux found in a number of detectors. Some physicists have even proclaimed that these messengers prove that WIMPs have been discovered. However, the story is a little more complicated.

Pairs of electrons ($e^-$) and their positron antipartners ($e^+$) are among the end products of WIMP annihilation. Although electrons are ubiquitous, positrons are more rare in the Universe. Thus observations of a large positron flux could be indicative of a region of high rates of dark matter annihilation. Protons in Earth's atmosphere produce a positron abundance that is a small fraction— about 1 in 10,000—of the number of electrons. This local abundance can be observed in laboratory measurements. But if WIMP annihilation is common in our Galaxy, the numbers from a region of high dark matter density could be much higher. To study positrons of galactic or cosmic origin requires detectors flying above the surface of Earth, such as balloons or satellites.

The HEAT (High Energy Antimatter Telescope) balloon experiment in the 1990s first discovered an anomalously large positron signal. Then in 2008 the Italian PAMELA satellite extended the measurements out to higher positron energies and found that the surplus persisted. Soon after, the FERMI satellite confirmed this positron excess. Further discussion of the FERMI satellite, primarily a gamma ray mission, will come later in the chapter.

The surplus positron signal created a burst of enthusiasm in the dark matter community. Researchers speculated that the excess could be attributed to annihilation of 200-GeV WIMPs (as a reminder, GeV stands for 1 billion electron volts, roughly the mass of a proton) in the Galaxy. Yet the data did not match the theoretical models without serious tweaking of parameters. To explain the high positron abundance seen, standard WIMP particles (such as those from SUSY) won't work. The signal needs to be enhanced by a *boost factor* of about 200. In other words, there must be some effect that increases the numbers of positrons coming from WIMPs by at least this much. One possibility is that Earth might be situated in an overdense region, or clump, of dark matter. But the results of computer simulations of galaxy formation, the best information we have about the locations of dark matter particles, do not support the existence of such clumps. Another way to get the boost factor is by means of an effect known as a Sommerfeld enhancement, a quantum mechanical effect that may apply to nonstandard WIMP particles. However, both these explanations are a stretch.

Much more likely is an astrophysical origin of the positron excess. The best explanation is due to pulsars, which are rapidly rotating neutron stars. As discussed in Chapter 5, neutron stars are the end-products of massive stars that explode as supernovae. Our Galaxy contains a multitude of these, including nearby Geminga, which exploded roughly 300,000 years ago. The FERMI satellite (which confirmed the HEAT and PAMELA positron excess) also found many new pulsars, leading to a better theoretical understanding of the objects. Ironically, this new information reinforced the neutron star explanation for the PAMELA excess rather than a dark matter explanation. Although there is as yet no consensus, many researchers find a pulsar origin for the positrons quite compelling. My bet is on pulsars.

In May 2011 the Space Shuttle Endeavour, the last flight in the Space Shuttle Program, ferried the Alpha Magnetic Spectrometer (AMS) up to the International Space Station. There the AMS detector is studying antimatter in the Universe. In April 2013, the AMS collaboration announced results that agree with the surplus of positrons seen in the other experiments. Unlike PAMELA, AMS can tell the difference between various types of particles in their data. Whereas the excess observed in PAMELA could have been due to misidentified antiprotons, AMS can differentiate these from positrons. AMS did more than confirm the excess seen by PAMELA; it proved it really exists.

AMS was also able to measure positrons with higher energies than any previous experiment. Theorists who believed in a dark matter interpretation of PAMELA data had hoped for a bump in the data, with a maximum count rate at a few hundred GeV followed by a decline at higher energies. Unfortunately for their theories, AMS found more and more positrons at higher and higher energies. Such a rise maintains the previous trend in the data but does not elucidate the origin of the positrons any better. Unfortunately, it will be difficult to distinguish conclusively among the different possible explanations for the excess.

A major problem with using positrons as tracers of dark matter is the uncertainties associated with their propagation. Positrons are positively charged particles and interact electromagnetically with other matter on their way to us. They are jostled about by these interactions and do not take a straight path. Because their tracks do not point directly back to their sources, scientists cannot tell where they were originally produced. As a consequence, the interpretation of the HEAT, PAMELA, and AMS data will remain uncertain.

### Neutrinos in IceCube/DeepCore at the South Pole

The IceCube/DeepCore detector at the South Pole is searching for astrophysical neutrinos from a variety of sources, including WIMP annihilation (Figure 8.7). An advantage of tracking neutrinos is that they have no electric

**FIGURE 8.7** The IceCube/DeepCore detector at the South Pole consists of kilometer-long strings of phototubes dug deep into the ice. The Eiffel Tower is shown for comparison. On the right is a postcard of a penguin I received from Antarctica. *(Right) pinguino k.*

charge. Unlike positrons, they undergo no interactions with other matter on their way to us. They travel in a straight line and thus are good tracers of dark matter. IceCube/DeepCore consists of kilometer-long strings of phototubes dug deep into the ice. Neutrinos interacting with the ice would light up the phototubes. This is one instance where I would badly like to be an experimentalist: whenever I receive another postcard of penguins sent to me by my friends at the South Pole, I think about how much I would like to get down there!

The best directions to look for WIMP annihilation neutrinos are toward regions of high WIMP density: the centers of Earth and the Sun (where WIMPs collect after being captured), the center of the Galaxy, and the dwarf spheroidal galaxies inside the Milky Way. Although there is as yet no signal in this detector, the future could be very interesting indeed.

If in fact the excess positron signal seen in PAMELA stems from WIMP annihilation, then significant numbers of neutrinos should be produced as well. As my collaborators and I showed in a paper in 2009, these accompanying neutrinos could be found in a matter of a few years by the IceCube/DeepCore detector at the South Pole.[16]

One of the best places to search for signatures of WIMP annihilation is toward the center of the Milky Way. Computer simulations indicate that the peak of the dark matter density is at the center of the Galaxy, with the abundance dropping rapidly farther out. The supermassive black hole near the Galactic Center may have drawn in yet more dark matter (without swallowing much of it) and enhanced the central abundance even further.[17] As a consequence of the large number of WIMP particles near the center, the annihilation rate would be significant, and many high-energy photons known as gamma rays may be produced. Such gamma rays are the most energetic form of light—even more energetic and biologically destructive than x-rays.

The large dark matter abundance at the Galactic Center could give rise to substantial annihilation signals. Unfortunately for WIMP hunters, a lot of competing astrophysical phenomena are taking place near the center, such as x-ray and other emissions from matter that is falling into the black hole. These effects make studies of the Galactic Center somewhat confusing, as it is hard to disentangle the physical origin of particles coming to us from there.

Nevertheless, observations of gamma rays from the Galactic Center provide an interesting probe of dark matter annihilation. Over the past decade one telescope after another has found anomalous events: EGRET, HESS, VERITAS, CACTUS, MAGIC, and others. Unexplained signals fueled hopes that WIMPs could be found via their gamma ray messengers.

The gamma ray observatory VERITAS is an array of four 12-meter telescopes. The original plan was to locate the array at Horseshoe Canyon on the way up to Kitt Peak National Observatory above the Sonoran Desert in Arizona. This location, one of the darkest night skies on Earth, has little contamination from light produced by human activity. However, after sightings of a Mexican Spotted Owl, a threatened species, the Bureau of Indian Affairs put a stop to the building of VERITAS at that location. After further legal battles, the collaboration moved the telescopes to a parking lot at the Fred Lawrence Whipple Observatory in Arizona and started taking data. As a consequence of the delay, the competing HESS experiment in Europe obtained results first, but VERITAS is now making interesting astronomical observations. When a gamma ray interacts with Earth's atmosphere, VERITAS observes a brief flash of light. This Earth-based array complements the FERMI satellite above the atmosphere. Because gamma rays don't make it through the atmosphere to the ground, direct measurements of the particles require a space-based observatory.

Currently the premier gamma ray mission is the Fermi Gamma Ray Space Telescope (FERMI for short), a joint venture of the U.S. Department of

Energy, NASA, and government agencies in France, Germany, Italy, Japan, and Sweden. The FERMI satellite was launched into low Earth orbit in 2008. On board FERMI is the Large Area Telescope. This telescope has 0.1 degree angular resolution, so that it can obtain detailed images of regions separated by as little as a tenth of a degree. With this resolution, it can take data from very specific directions of interest, including the center of the Galaxy.

The FERMI data are publicly available, so that anyone can scour them for interesting results. Based on the data, a series of claims have been made about possible WIMP origins of the signal. Douglas Finkbeiner, formerly a University of Michigan undergraduate student (one of my favorite cosmology students) and now Professor of Astronomy at Harvard, studied the data taken by FERMI in its first year of observations. He found an excess gamma ray signal in the inner Galaxy that he dubbed the FERMI Haze. Previously he had discovered a similar excess of microwaves in WMAP data (the WMAP Haze). At first he thought both signals could be caused by dark matter annihilation. However, as more data came in, a reanalysis showed a more complicated story. The excess gamma rays come from a region with a spatial shape now called the Fermi Bubble, which resembles two bubbles on either side of the Galactic plane meeting in the center. The outer parts of these regions are likely caused by astrophysical effects unrelated to WIMPs. Dan Hooper, Lisa Goodenough, Tim Linden, and Tracy Slatyer, however, argue that the gamma rays from the innermost parts of the bubbles could be from annihilation of 10-GeV WIMPs. Here too, just as in the DAMA, CoGeNT, CRESST, and CDMS laboratory experiments, the 10-GeV WIMP story is heating up.

Another good direction for FERMI's gamma ray searches is toward dwarf galaxies. These are small galaxies with masses ranging from a millionth to a thousandth of that of the Milky Way. There are roughly two dozen known dwarf galaxies in our Galaxy, remnant substructures of its formation process. As discussed in Chapter 2, the Galaxy grew by the mergers of such smaller objects, and some of them remain intact today. Dwarf galaxies are excellent places to search for WIMP annihilation signals for several reasons. First, they are among the most dark matter–dominated objects in the Universe. In addition, there is no competing astrophysical gamma ray production expected, unlike the case of the Galactic Center. As yet no excess gamma ray emission has been found. The California Institute of Technology's Jennifer Siegal-Gaskins, convener of FERMI's dark matter group in 2012, reported limits on dark matter annihilation from a combined analysis of the null signals from 10 dwarf galaxy targets. The current results are just beginning to provide interesting tests of the favored WIMP candidates. (Ironically, further analysis a year later actually led to weaker bounds). With increased observation time and the discovery of new dwarfs, the projected future bounds on dark matter could become very stringent or signs of dark matter will become apparent.

For dark matter hunters, an exciting result from the FERMI satellite was the 130-GeV gamma ray line coming from near the Galactic Center. This is a signal that could arise from the annihilation of two WIMPs directly to two photons. Typical gamma rays from WIMP annihilation are the final products of a long decay chain. For example, WIMPs might annihilate to quarks and antiquarks, which then pick up companion quarks to make pions. The pions decay to a variety of particles in succession. At the end of the chain, gamma rays are produced with a wide range of energies, typically comparable to a tenth of the energy of the original WIMP mass. Such a broad signal is in general difficult to distinguish from background.

In contrast, if two WIMPs annihilated directly to two photons without any intermediate steps, then the energy of each of the outgoing photons would be exactly equal to the mass of the incoming WIMP. This is known as a *monoenergetic line*. In the 1980s, Lars Bergström from Stockholm University pointed out that the discovery of such a line could be powerful evidence for WIMP detection. Astrophysical sources not related to dark matter are unlikely to produce gamma ray lines. Currently Bergström is director of the Oskar Klein Centre for Cosmoparticle Physics in Stockholm as well as a member of the FERMI team.[18]

In 2012 Christoph Weniger, a postdoctoral fellow at the Max Planck Institute in Munich, scrutinized the FERMI data and found preliminary evidence for a line feature at 130 GeV. If confirmed, this signal would be consistent with a 130-GeV WIMP. Weniger's potential discovery generated a flurry of excitement and hundreds of papers within a few months. However, the signal is getting weaker with time. Once the FERMI team reprocessed and reanalyzed the data, its statistical significance fell. Yet some features of this 130-GeV line data are still unexplained. As Doug Finkbeiner says, "I am in the position of saying the signal is not behaving well (and is probably wrong), but it is wrong in a rather mysterious way. I think it is important to keep on this until we either understand what is wrong with the instrument/data reduction, or understand something about physics."

### The Arrogant Frog

While I am on the subject of science in Stockholm, I can't resist a story about an amazing ceremony I took part in. In September 2012 I was invited to Sweden to receive an honorary doctorate (honoris causa) from Stockholm University. Physics professor Lars Bergström had nominated me for my work as "one of the world's best known astroparticle physicists." (I don't mean to sound like "the arrogant Freese." This phrase was written by Lars, not by me!) For this traditional and formal occasion, I bought a ballgown at Saks Fifth Avenue in New York. It was a most remarkable event. The night before the actual ceremony there was a party in the "Ghost Castle," a mansion originally owned by a wealthy German industrialist who had accumulated a remarkable art collection. At the party we enjoyed looking at the masterpieces displayed on the walls. After his death the industrialist donated the building to the University. Rumor has it that his ghost still wanders the rooms of the mansion.

The ceremony itself was held in City Hall, following the same tradition as the Nobel Prize. The main brick-walled room seats thousands of people and has ceilings that are two stories high. There were four of us who received honorary degrees for research in the natural sciences. One by one we walked up a staircase to a platform, where we each received a diploma, an inscribed gold ring, and a laurel wreath. Since Saks Fifth Avenue had forgotten to mail the jacket that came with the gown, I wore a purple scarf around my shoulders instead. On the way back down the staircase, I had to hold onto the scarf, the ring, and the diploma, and lift the ballgown off the floor, all while ensuring the laurel wreath stayed on my head. I was terrified that I would trip, but I didn't. An elegant dinner for 850 people followed in a ballroom with walls decorated with gold mosaics. The waiters sang as they brought the food on candlelit trays. We were served French wine from the vineyard "The Arrogant Frog." Dinner was followed by a dance in the main hall with a band playing songs ranging from Sinatra to disco tunes. The ceremony for the honorary doctorate in Stockholm is probably the best one in the world, and I was honored to be part of this event.

### The Future of Dark Matter Experiments
Huge Vats of Gas or Small Amounts of DNA

A major step forward in the field of dark matter detection would be the development of detectors with directional capability—the ability to determine which direction the detected WIMP came from. With this information, it would be much easier to prove WIMP discovery. Current direct detection experiments require thousands of WIMP events to be conclusive. In contrast, with directional information available, as few as 30–100 counts from WIMP interactions would suffice.

When a WIMP scatters off of a nucleus in the detector, the nucleus typically gets kicked in the forward direction. Thus if scientists could determine the track of the nucleus, they could extrapolate backward and identify the path of the incoming WIMP.

Directional sensitivity would be a coup, because it would immediately allow measurement of the "head/tail" asymmetry of the incoming WIMPs. As described earlier in the chapter, because of the motion of the Sun around the center of the Milky Way, we are moving into a wind of WIMPs. As a consequence, 10 times as many events are expected when looking into the wind as in the opposite direction. There are absolutely no spurious signals that would be expected to exhibit this difference. The trick would be to compare the count rates 180 degrees apart. If researchers could count the signals into and against the WIMP wind, fewer than 100 events in the detector would be enough to argue that WIMPs really have been discovered.

Once the head/tail asymmetry is measured, the second generation of directional detectors should be able to detect an even more convincing signature of WIMPs: the daily modulation of the event rate stemming from the rotation of Earth. The direction of the WIMPs changes with respect to detectors on Earth by about 90 degrees every 12 hours. Earth's rotation causes the WIMP count rate to vary significantly with the time of day in a highly predictable way.

An experiment that could measure both the annual and daily variation of the signals would be a surefire test of WIMP detection. Measurement of the diurnal modulation would determine the direction of the WIMP wind, which could then be compared for consistency with an annual modulation signal found in a different experiment. Another feature of directional detectors would be their ability to study galactic substructure. Streams of dark matter sweeping through the Galaxy would show up as spikes coming from one particular direction.

### Huge Vats of Gas

Experimental efforts with the goal of directional dark matter detection are taking two different approaches. The aim is to record the tracks of recoiling nuclei from WIMP collisions, because these move in essentially the same direction as the incoming WIMP. One technique has been slowly developing over the past two decades and uses huge vats of gaseous detectors.

Traditional dark matter detectors have a severe limitation: the nuclear recoil tracks are shorter than the resolution capabilities of the detector. The nuclei stop within 10 nanometers (a nanometer is a billionth of a meter) of being hit by WIMPs, whereas existing detectors have spatial resolutions 100 times longer—a few microns. There is no way to identify the nuclear track. To get around this problem, scientists are building gas detectors, where the density of target atoms is much lower than in a solid or a liquid. In this case the recoiling nuclei move much farther before they are stopped. The particle tracks are long enough to be measured with conventional techniques.

To get to the low densities of the gaseous phase, scientists pump the material down to low pressure, perhaps a tenth of atmospheric pressure. Then the problem is that the gas is so diffuse that WIMPs rarely interact at all. To combat this issue, there must be a huge amount of gaseous target material. The detector would have to weigh more than a metric ton and occupy a volume of 10,000 cubic meters, filling an entire cavern in an underground laboratory. This can be done, with careful shielding against the radioactive backgrounds from the surrounding laboratory. So far researchers have constructed smaller prototypes made of low pressure $CF_4$ gas. The DRIFT detector currently operates with 30 grams in a volume of 1 cubic meter, whereas the DMTPC detector has 3 grams of target material. The two collaborations aim to scale up to much larger masses.

### Dark Matter Detectors Using DNA

Recently I have become involved in an exciting project to develop a novel type of directional dark matter detector—a smaller, cheaper alternative. Our plan is to build dark matter detectors using DNA. The goal is to determine the track of a recoiling nucleus from a WIMP collision with an accuracy of a nanometer. This is the typical distance between nucleotides (A, G, C, T) in a strand of DNA. Our proposed technique is a way to use this length scale to build particle trackers with nanometer resolution.

The idea originated with Andrzej Drukier, the same person I worked with in the 1980s on the papers proposing dark matter annual modulation. In 1990 Andrzej left physics, started a biotech company, and worked in the field of biology for 20 years. During that period he befriended many well-known biologists. In 2010 he resurfaced in the dark matter community and started talking about using biological techniques developed over the past two decades as tools for dark matter detection. David Spergel (the third of the "Three Musketeers" from the early dark matter papers), Andrzej, and I, together with three distinguished biologists, are coauthors of the initial paper on the new idea for dark matter detectors using DNA.

Andrzej's eyes always twinkle as he smiles through his bushy beard. Because he is Polish (and worked on astrophysical searches for magnetic monopoles) I call him the monopole. He calls me "Dear Lady" and describes himself as "a very conservative anarchist with hippy tendencies." Back in Poland he got into trouble for resisting the communist regime, though he also dated the daughter of the head of the second largest party in Poland. In a picture I took of Andrzej at a restaurant in Los Angeles, I realized he looks like a rabbi. He then said, "Once upon a time I was called rabbi. Here in U.S., I've become much more normal, much more tough."

In the spring of 2012, Andrzej visited me at the California Institute of Technology in Pasadena, where I was on sabbatical. (Every 7 years, faculty members are given a year to dedicate to research, away from teaching and committees. The goal is to stimulate new thinking and new projects.) Andrzej and I went to San Diego to talk to his friend Charles Cantor, one of the originators of the ideas for the Human Genome Project and founder of the company Sequenom. Apparently his private venture has been highly successful. On the way into his house we spied a Maserati in his garage (I'm a huge fan of fast cars). We sat by the pool in Charles's backyard together with Takeshi Sano, another great biologist. I was amazed that Charles not only took an interest in these ideas of dark matter detectors with DNA but also indicated that the biological techniques required were doable—even commonplace. The devil of course is in the details. The killer in the end may be the radioactive backgrounds, as usual in this business. Andrzej was full of creative (almost crazy) ideas, and together we narrowed it down to a project that could really be built. Another major player in the field of genetics, George Church from Harvard University, rounded out our group of six.

After much discussion, we decided on a *DNA tracking chamber* design. The detector has two components: a target material for WIMP interactions and quadrillions of DNA strands. The basic idea is that a WIMP strikes a nucleus in the target material and knocks the nucleus into a periodic array of single-stranded DNA. The nucleus breaks any DNA strands that it hits. We can identify the location of the breaks using well-known biological techniques. Looking at the trail of broken DNA, we can reconstruct the path of the nucleus. Because the WIMP is typically moving in the same direction as the nucleus, we can determine the path of the incoming dark matter particle.

The simplest implementation consists of thin foils of gold as the target material, about 5 nanometers in thickness (Figure 8.8). Hanging down from the gold foils are strings of single-stranded DNA—like a series of beaded curtains. The DNA strands all consist of identical sequences of bases (combinations of A, C, G, and T nucleotides) roughly a nanometer apart, with an order that is well known. An incoming WIMP from the halo of our Galaxy strikes one of the gold nuclei and knocks it out of the gold plane and into the hanging DNA. The gold nucleus traverses a few hundred DNA strands before stopping. When the gold hits a DNA strand, the DNA breaks, and the cutoff segment falls down onto a capture foil, where it is periodically removed. The locations of

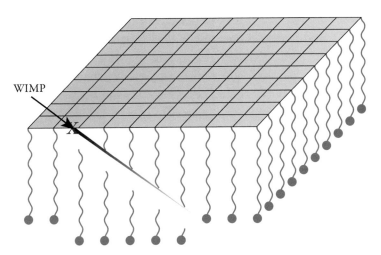

WIMP

**FIGURE 8.8** A novel idea for dark matter detectors using DNA. A Weakly Interacting Massive Particle (WIMP) from the Galaxy hits a thin foil of gold and knocks a gold nucleus into hanging strands of DNA. The gold severs any DNA strand it hits. Biologists can identify the location where the DNA broke off and reconstruct the track of the recoiling gold nucleus (with nanometer accuracy). Because the WIMP scatters the gold in the forward direction, this technique identifies the path of the incoming WIMP. *From A. Drukier, K. Freese, D. Spergel, C. Cantor, G. Church, and T. Sano. "New Dark Matter Detectors Using DNA for Nanometer Tracking." arXiv:1206.6809 [astro-ph.IM].*

the breaks in the DNA strands are easy to identify using techniques that have been refined over the past 20 years. The technique of polymerase chain reactions (PCR) makes a billion copies of the broken segment and gene sequencing determines the order of the nucleotides. In the words of Jeff Filippini of the CDMS experiment, "Errors in DNA are easy to identify and to replicate." Combining the information from all the broken segments, we can track the path of the recoiling nucleus with nanometer accuracy.

The DNA tracker will consist of thousands of modules—each with a layer of gold foil with hanging DNA strands—separated by sheets of mylar. As a nice image for the device, the actor Sarah Kennedy Flug envisioned dangling earrings (though here the gold is attached to the earrings).[19] The first round of the experiment will require 1 kilogram of gold and 0.1 kilogram of DNA. The total mass is much less than the thousands of kilograms proposed for other conventional detectors. Surprisingly, it is the DNA that is more expensive, not the gold!

Like all other dark matter detectors, the DNA tracker will be plagued by false background signals and must operate deep underground. Radioactive contamination from nearby rock or from the detector itself can be a serious problem. One advantage of the DNA tracker design is that WIMPs, which stop in only one gold/DNA layer, can be differentiated from the more energetic background particles, which should traverse multiple layers. Another advantage is that the DNA tracker can operate at room temperature, whereas many other dark matter detectors must be cooled to very low temperatures, and the cooling mechanisms tend to introduce radioactivity.

Commercially available off-the-shelf modules are currently available for a few hundred dollars. Each module consists of a layer of gold attached to aperiodic arrays of single-stranded DNA that are 250 bases in length. We can use these simple prototypes to perform some initial tests. In the long run we need radioactively clean, longer strands in a periodic array. Natural DNA (such as in the human body) has two sources of radioactivity: carbon-14 and potassium. For the experiment, we need to manufacture DNA with ancient carbon (so that the carbon-14 has had time to decay) and with potassium replaced by a different element. Harvard geneticist George Church is developing pure and straightened DNA strands.

The first goal will be to search for a head/tail asymmetry caused by WIMP interactions. The count rate in the direction of the WIMP wind should be 10 times higher than in the opposite direction. Our design automatically provides this head/tail distinction. It is the gold that breaks the DNA, not the WIMPs. The WIMPs have to propel the gold nuclei in the direction of the DNA curtain to get any detectable events. WIMPs that go in the wrong direction, traversing first the DNA and then the gold, will knock the gold nuclei into the mylar instead, not producing any signal. The maximum count rate should be when the WIMP wind reaches the gold first and then the DNA; the minimum count rate would be the opposite, when the wind is directed first into the DNA and then the gold. By rotating the detector 180 degrees, the signal should change by a factor of 10. Once this head/tail asymmetry is found, the goal of the next generation detectors will be to look for the actual track of the recoiling nucleus with nanometer resolution, using longer DNA strands in a periodic array.

This novel approach of detecting dark matter with DNA is still in its infancy. We need to do many tests and calibrations to see whether such a detector can be made to work. The University of Michigan recently initiated the MCubed program with the mission of jumpstarting interdisciplinary new ideas, such as ours. To be eligible for this program, groups of researchers must involve at least three different departments in the university. With the seed funding we obtained from MCubed, we are now performing initial tests of the DNA tracker.[20] In the long run, there could be a variety of other applications of nanometer trackers.

Although the details of the design may yet change drastically, the basic idea of dark matter detection using DNA is a method for the future.

There is a beautiful symmetry in these scientific developments. Over the past decades, studies of element abundances from Big Bang nucleosynthesis established the need for nonatomic dark matter in the Universe. The elements that were produced in stars—carbon, nitrogen, oxygen, and phosphorus—are the building blocks of the DNA and RNA that are the fundamental basis for all life. Now we have come full circle. It is an intriguing prospect that we can use the DNA created from stardust to search for astrophysical dark matter particles.

## Upcoming Laboratory Dark Matter Searches

Laboratory dark matter experiments are poised at an exciting juncture. Numerous collaborations worldwide are continuing direct detection searches for dark matter, including ANAIS, ArDM, CDEX/TEXONO, CDMS, CoGeNT, COUPP, CRESST, DAMA, DARKSIDE, DEAP/CLEAN, DM-Ice, DRIFT, EDELWEISS, EURECA, KIMS, LUX, NAIAD, PandaX, PICASSO, ROSEBUD, SABRE, SIMPLE, TEXONO, XENON, XMASS, and WArP.

One direction for the future of the field is the scaling up of existing detectors in mass and size. Experimentalists are building liquid xenon, argon, and other detectors that weigh more than 1 metric ton, 10 times bigger than any current detectors. Such experiments should probe some of the most promising ranges of WIMP mass and interaction strength, including those where SUSY dark matter is expected to exist. The United States has at least three different teams that are competing for government grants for a scaled-up version of the current XENON experiment. Elena Aprile proposes to build up her current XENON detector. Rick Gaitskell and Tom Shutt lead the LUX effort. A new competitor is PandaX. The U.S.-based groups are planning similar experimental efforts and will likely merge their collaborations.

Direct detection experiments search for three main experimental signatures for WIMP detection: phonons, charge, or light. All three can be produced by WIMP collisions with nuclei. Experiments operating today search for one or two of these signals. Phonons are basically heat; charge is created as WIMPs ionize the detector medium; and photons are the result of a target converting WIMP energy to scintillation light.

Some experiments only measure one of these three signals. DAMA and KIMS only use scintillation light, and CoGeNT only measures charge. Yet background rejection is far more powerful when two of these three detection methods are used. For example, the liquid xenon detectors measure light and charge; CDMS and EDELWEISS measure heat (phonons) and charge. To study the 10-GeV WIMPs of current interest, which correspond to lower energies in the detector than the original design goals, some experiments have been forced to abandon one of their two methods, making the results less reliable. Experimentalists are rethinking approaches that allow them to attack this low-mass regime.

New creative experimental designs are being developed as well. Reina Murayama of the University of Wisconsin has started an exciting new program called DM-Ice. Located at the South Pole, this experiment embeds sodium iodide (NaI) crystals deep into the ice, as far as a kilometer below the surface. The dectector is made of the same material as the DAMA experiment and should be able to test the current DAMA claims of WIMP detection. Because DM-Ice is in the Southern Hemisphere, it offers a great comparison point for DAMA's location in Italy. If the same annual modulation of the signal is seen in both places, it will provide a persuasive argument that WIMPs have been found. The location in Antarctica has some clear advantages. Little natural temperature variability occurs beneath the ice there, and thus no reason exists for yearly variation of the signal other than because of dark matter. The ice at the Pole is incredibly pure, with a flux of neutrons from radioactivity much lower than in any underground facility. The DM-Ice group is very impressive. From the time Murayama first started pushing the idea, it took only 6 months to deploy a small amount of detector material and make it work. The group obtained the crystals currently used from the NAIAD experiment in the United Kingdom. These are not clean enough, nor are there enough of them. The next challenge for DM-Ice will be to obtain crystals comparable to those DAMA has. Because the DAMA group has an exclusive contract with the manufacturer of the crystals they are using, DM-Ice is pursuing other options.

The advance in dark matter experimental sensitivity has been impressive. The progress over the past 25 years has been similar to Moore's law in the field of computer science. In the semiconductor industry, the number of components on integrated circuits, processing speed, and memory capacity all roughly double every 2 years. In dark matter experiments, the sensitivity has improved even more quickly, by about a factor of 10 every 2 or 3 years. This pattern can be expected to continue for a while, but not indefinitely. In the long run, there is a lower cutoff to the WIMP-nucleus interaction strength that can be probed. Eventually, the signal will be swamped by astrophysical neutrinos from the Sun, stellar explosions, and other sources. At low enough sensitivity, these neutrinos will start to dominate what the detectors will see. Even then the neutrinos and the WIMPs might still be differentiated, because the energies they deposit in detectors at different energies will not be the same—but finding WIMPs will become more and more difficult. The neutrino physics that can be studied will be fascinating, but that is another subject entirely. In the meantime, the next decade should be exceptionally exciting for dark matter hunters.

How Can We Be Sure?

What will it take for the physics community to agree that a discovery of dark matter particles is conclusive? Multiple experiments need to detect compatible signals. One possibility would be for several direct detection experiments in search of astrophysical WIMPs striking laboratory nuclei to find consistent results. For example, DAMA is made of sodium iodide, CDMS of germanium, and XENON of xenon. If the data from all three groups pointed to the same particle, the convergence would be highly persuasive. Further, by comparing results from the different detector materials, scientists could determine WIMP properties, including the mass and interaction strength.

A "smoking gun" in astrophysical searches would be detection of both diurnal and annual modulations. Compatibility of the daily and seasonal variations of the signal with the expected direction of the WIMP wind would be strong evidence for the existence of WIMPs.

Another path to definitive discovery is indirect detection of a variety of dark matter annihilation signals, again in multiple experiments. Dark matter can annihilate through numerous channels to gamma rays, neutrinos, and positrons. One could imagine detection of several of these signals in different experiments, all arriving from the same source in the sky. Then it would be likely that their origin is the annihilation of dark matter particles, and a discovery would be confirmed.

Despite the uncertainties, theorists have become tremendously excited by all the recent hints of detection and are trying to reconcile the apparent discrepancies among the data. Some theorists are proposing exotic models involving a "dark sector" of the Universe containing particles that interact with us only indirectly. Some speculate that the dark matter lives on another *brane* (short for membrane), or 3-dimensional surface parallel to the one we inhabit, situated in the larger 10-dimensional Universe of string theory. Perhaps there is more than one type of dark matter particle. These theoretical debates will resolve with the unambiguous detection of dark matter particles and determination of their properties.

The experimental side of this field is booming. In the past few years, the sensitivity has improved by a factor of more than 100. In the next few years, we can expect this trend to continue. Soon there will be more data from both direct and indirect detectors as well as from the LHC at CERN in Geneva. The future of direct detection experiments over the next decade includes enormous vats of noble liquids, containing up to 20 metric tons of xenon, neon, or argon. In addition to scaling up the current detectors, new technologies are being developed, such as gaseous detectors and a nanometer tracking device using DNA. The answer to the dark matter puzzle is near.

# NINE

## Dark Energy and the Fate of the Universe

> Don't let the bright lights fool you.
> The Dark Side controls the Universe!
> Dark matter holds it together.
> Dark energy controls its destiny.
>
> GREG TARLE,
> professor at the University of Michigan

As strange as the concept of a new type of dark matter particle may be, the concept of dark energy is yet more bizarre. All matter, including atoms and dark matter, amounts to only roughly a third of the total content of the Universe. The remaining two-thirds of creation consists of the mysterious dark energy. We know of its existence from studies of supernovae, the bright explosions of dying stars. In 1998 two independent groups of astronomers observing distant supernovae found that they were significantly fainter than expected. The reason, the scientists postulated, was that the supernovae are accelerating away from us.

By definition, all matter (atomic and dark matter) feels the force of gravity. Mass of any kind is attracted to mass of any other kind. It is this attractive aspect of gravity that holds galaxies and clusters together, as well as prevents our bodies from floating off Earth and into space. In contrast, dark energy is in some sense antigravitating, effectively producing a repulsive force that causes every point in the Universe to accelerate away from every other point. Scientists have known since the 1920s that the Universe is expanding, but the discovery that the expansion is speeding up came as a huge surprise.

The name "dark energy" is a bit of a misnomer, in that there is not necessarily any relationship between dark matter and dark energy. Whereas matter and energy can be transmuted into one another, the same is not true of dark matter and dark energy. In this sense their names give a false sense of connection between the two quantities. They do share the fact they don't give off light and thus can't be observed in telescopes. They are both components of the "Dark Side" of the Universe.

### Rounding Out the Universe: Type IA Supernovae and Dark Energy

The completely unexpected discovery of dark energy created a paradigm shift in cosmology. Not only is its nature a complete mystery, its existence also defies any reasonable theoretical expectations. Several proposals to explain its origin have been made. In each of them, the changes to the standard theories are radical. This chapter describes the observational evidence for dark energy from studies of supernova explosions and explores possible theoretical explanations.

### Supernovae as Standard Candles

To study the expansion history of the Universe, astronomers use *standard candles,* objects that shine with the same luminosity, or brightness, no matter at which epoch they exist. A telescope looking far away amounts to one looking far back in time, because the information comes to us at the speed of light; by looking at distant objects, astronomers can learn about early times. The comparison of nearby and distant objects is then meaningful only if they shine equally brightly at any epoch. Most stars and galaxies are not good standard candles, because they evolve with time. Galaxies that existed 10 million years ago look completely different from galaxies that exist now. Currently the best standard candles are Type IA supernovae, bright explosions of dying stars that are well understood and are equally luminous no matter where or when they form.

Supernovae, which include some of the brightest, most beautiful objects in the sky, are the explosions of self-destructing massive stars (Figure 9.1). They arise from explosive events that take place at the ends of stellar lifetimes. A supernova can radiate as much energy in the brief time span of a few weeks as the Sun emits over its entire lifetime.

A variety of mechanisms for supernova explosions exist, but only the Type IA supernovae are uniform enough to be used as standard candles. These are the bright explosions of white dwarfs merging with other stars, either ordinary stars or other white dwarfs. Although there are only one or two of these explosions per millennium per galaxy, they are as bright as ten billion suns and can be seen at cosmological distances.

As explained in Chapter 5, white dwarfs are the stellar remnants of stars that were once about as massive as the Sun. Five billion years from now, when our Sun runs out of nuclear fuel, it will puff up into a red giant large enough to engulf the inner planets, including Earth. A million years later the red giant will collapse to a cold, faintly glowing white dwarf about the size of Earth. The Milky Way Galaxy alone may contain millions of these white dwarf stars.

Half of all stars in the Universe are in binary systems, in which two stars orbit around each other. When a white dwarf is in a binary system with a com-

**FIGURE 9.1** (A color version of this figure is included in the insert following page 82.) A supernova remnant. *(X-ray) NASA / CXC / SSC / J. Keohane et al.; (infrared) Caltech / SSC / J. Rho and T. Jarrett.*

panion star, the stage is set for a Type IA supernova. The white dwarf begins to pull mass off the other star into an accretion disk of swirling matter. The white dwarf becomes more and more massive at the expense of its companion. Eventually it reaches a limiting value of 1.4 times the mass of the Sun. This is the maximum mass a white dwarf can sustain and is known as the Chandrasekhar limit.

Physicist Subramanyan Chandrasekhar first proposed the existence of white dwarf stars and recognized that they could never exceed this limiting mass. We've seen in Chapter 5 that the pressure support for a white dwarf is based on the Pauli exclusion principle from quantum mechanics. According to this principle, no two electrons can exist in identical states. When the mass of a star is condensed into a white dwarf, the electrons push back against being squeezed any further. Once the white dwarf mass approaches the critical value of 1.4 times the mass of the Sun, the electron degeneracy pressure supporting the white dwarf fails. The temperature at the center of the white dwarf approaches a few billion degrees, thermonuclear instability sets in, and the white dwarf explodes. Such an exploding white dwarf is known as a Type IA supernova. A powerful shock wave moves outward into the interstellar medium and sweeps up the surrounding gas and dust into an expanding shell known as a supernova remnant. The expanding remnant shines in a number of different wavelengths and produces extraordinarily beautiful images.

An exploding Type IA supernova is a short-lived but extremely bright event. Initially, it is 5 billion times as bright as the Sun. Then its luminosity tapers off in a predictable fashion over the course of a month. The term "light curve" is used to describe this changing light output over time. Because the explosions take place when the white dwarf mass approaches the Chandrasekhar limit, the light curves should be the same for all Type IA supernovae (once the correction factors in the next paragraph are accounted for). The consistency of the light output is due to the run-up to a nearly identical predetonation white dwarf. For this reason, Type IA supernovae make excellent standard candles. Astronomers can trust that they know the exact brightness of the explosion, no matter how distant (or how far back in time) the event took place.

Because of the differing compositions of white dwarfs, supernova light curves do vary. Fortunately astronomers can measure the element abundances and correct for this effect. They determine the metal content by observing spectral lines at specific wavelengths characteristic of the different metals.[1] Once the composition is known, the light curve of the Type IA supernova is predictable. A second adjustment is known as the *stretch correction*. The peak brightness of different Type IAs can vary by as much as a factor of two. Luckily astronomers can easily account for this variation by comparing the shapes of the light curves. The brightest supernovae also last the longest and have the broadest light curves. Brighter is broader. In contrast, the least-luminous supernovae have the steepest declines. With one simple corrective factor, scientists can shift the curves to line up appropriately. With these two corrections, the errors obtained by assuming Type IAs are standard candles become very small.

Standard candles in the Universe can be used to measure distances. The idea is to compare the known brightness (the *absolute luminosity*) emitted by the object with the amount of light that reaches our telescopes (the *apparent luminosity*). The farther away the object is, the dimmer it will appear to be. The apparent luminosity decreases with the square of the distance to the light source, so that by comparing the absolute and apparent luminosities, astronomers can ascertain the distance to the object. As an analogy, consider a light bulb. Imagine a large auditorium with a lecturer at the front of the room. If the speaker holds up a shining light bulb, it will look substantially dimmer to the people in the back row. The light beam has to travel a fair distance, and it spreads out en route. We know the actual intensity of the light bulb, because we can read the wattage on the bulb. For example, it could say 60 watts. Then, by comparing the emitted light intensity with how bright the bulb looks to the person in the back row, we can calculate the distance from the front of the auditorium to the back.

Similarly, astronomers use their knowledge of the intrinsic brightness of a supernova to estimate its distance from us. They know how bright it really is and can measure how bright it looks here on Earth. Assuming the Type IA supernova is a standard candle, they can compare these two intensities to find the distance to the object. The quantity obtained in this way is known as the *luminosity distance.*

Type IA supernovae are the only known objects in the Universe that are of equal intrinsic brightness regardless of when they occurred—billions of years ago or today. For this reason, they make excellent cosmological probes of early times. Although they glow brightly for only a few months, their light curves are distinctive. In the 1970s astronomers Stirling Colgate and Gustav Tammann realized the power of these supernovae as tools for studying the Cosmos. They argued that as few as 25 ancient supernovae, from at least 6 billion years ago, would be enough to determine the early expansion history of the Universe and to disentangle deceleration from acceleration.

## Accelerating Universe

For decades astronomers searched for enough Type IA supernovae to determine the deceleration or acceleration of the Universe's expansion. However, the observations were difficult. To quote Bob Kirshner of Harvard's Astronomy Department, "the path was not smooth or straight."[2] Finding enough supernovae to have a statistically significant sample was arduous. Dust absorption and other effects were hard to disentangle from true cosmic evolution. In the mid-1990s technological developments including better CCD cameras (similar to everyday digital cameras but much more sensitive) improved the ability to discover distant supernova events.[3]

In the meantime, Ruth Daly, then at Princeton, used a different technique. She studied radio galaxies, which don't qualify as standard candles but instead have well-understood sizes that could also be used to determine the expansion of the Universe. In January 1998, she put out a press release in which she stated that she had found that the most distant radio galaxies were extremely large, consistent with an accelerating Universe. Yet until the results of the supernova studies weighed in, the community was not convinced.

In 1998 two groups, the High-$z$ Supernova Search Team and the Supernova Cosmology Project, obtained light curves for dozens of Type IA supernovae from as far back in time as 8 billion years ago. The two observing teams simultaneously made the same discovery.[4] The ancient supernovae are 20% dimmer than had been expected. The implication is that they are farther away than predicted by the ordinary expansion of Hubble's law. The revolutionary conclusion is that the supernovae are accelerating away from us, because they are participating in an overall acceleration of the Universe.

Both groups of astronomers plotted the supernovae in their data in terms of a distance-versus-redshift diagram—effectively the distances to the supernovae versus the times of their explosions. Figure 9.2 shows their data points. In 1929 Hubble had made a similar analysis of distant galaxies. He had found that they followed a straight line, and he concluded that the Universe is expanding. The reason for using the Type IA supernovae as standard candles was to look farther back in time to see whether there is a deviation from this straight line. If the curve tilts downward (as had been expected), then the Universe is decelerating; if it tilts upward, then the Universe is accelerating.

In Figure 9.2, the vertical axis is the measured brightness of the supernova. As discussed above, from this quantity astronomers can determine the (luminosity) distance to the explosion. The conversion relies on the fact that the Type IAs are standard candles. A comparison of the known intrinsic light from the supernova with its apparent brightness leads to an accurate distance measurement of the object.

The horizontal axis dates the supernova. The quantity shown is the redshift, a measure of the age of the Universe when any event took place—in this case the supernova explosion. Specifically, the redshift tells us how much the Universe has grown since the time of the event. Light can be thought of as a wave, and the length of the wave stretches as the Universe expands. Astronomers observe spectral lines at specific wavelengths characteristic of different metals in the galaxy hosting the supernova. By measuring how far these wavelengths have stretched, scientists can deduce when the supernova took place.[5] The numerical value of the redshift dates the supernova explosion. For example, an explosion at $z = 1$ happened almost 8 billion years ago, an explosion at $z = 0.1$ happened 1.3 billion years ago, and one at $z = 0.01$ happened 140 million years ago.

Figure 9.2 effectively plots the distance to a Type IA supernova versus the time of its explosion. A car traveling along a highway provides a good analogy. A similar plot would show the distance traveled by a car versus the amount of time it took (for example, 60 miles in 1 hour). The slope of the line would then be 60 miles per hour, the speed of the car. A straight line would correspond to a constant speed, whereas a curve tilting upward would imply acceleration. Similarly, the slope of the line in the redshift versus distance plot indicates the speed of the supernovae. A straight line would correspond to constant speed; an upward tilt in the curve would imply acceleration.

The two groups of astronomers, the High-$z$ Supernova Search Team and the Supernova Cosmology Project, were surprised by their results. The curve from their data shown in Figure 9.2 slopes upward, away from a straight line. This tilt implies cosmic acceleration. To explain the data, the researchers had to add a new dark energy component to the Universe.

Theorists had previously contemplated a variety of hypothetical universes made of different constituents. Cosmologists define $\Omega_M$ to be the fraction of the Universe made of matter. This quantity contains the sum total of atomic matter plus dark matter. They also define $\Omega_\Lambda$ to be the fraction of the Universe in the new component, the dark energy. In Figure 9.2 the solid curves correspond to different hypothetical universes with different amounts of these two ingredients. Among the possibilities imagined by theorists were a flat Universe made only of matter ($\Omega_M = 1$); a flat Universe containing both matter and dark energy ($\Omega_M = 0.3$ and $\Omega_\Lambda = 0.7$); and a spherical Universe containing only matter ($\Omega_M = 0.3$ and $\Omega_\Lambda = 0.0$).

The new data allowed the astronomers to test the models. The dots (with error bars) in Figure 9.2 indicate the data from the Type IA supernovae. Although it is difficult to disentangle by eye, statistical analysis shows that a Universe consisting only of matter simply does not match the data. Instead, the best match between data and theory requires that the Universe has (roughly) $\Omega_M = 0.3$ and $\Omega_\Lambda = 0.7$.

The cosmology community was incredibly excited when these results were released. Yet the excitement was mixed with a healthy amount of skepticism. One concern was that dust between the supernovae and the Earth could obscure some of the light and cause the stars to look fainter for mundane reasons. Another was the question of how uniform the supernova light curves really are: in reality, how perfect are the Type IAs as standard candles?

I, for one, was unsure at first how seriously to take these results. I must admit that I was biased by my theoretical prejudices. As described later in the chapter, the implications of an accelerating Universe for cosmological theories

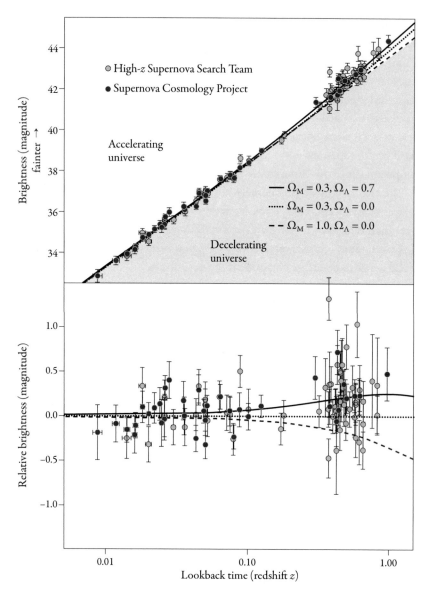

**FIGURE 9.2** The supernova data that proved that the Universe is accelerating. The upper panel shows the brightness of Type IA supernovae at different redshifts, or lookback times. Effectively this is a plot of the distance to the supernovae versus the time of the explosions. The 1998 data (dots with error bars) were taken by the High-z Supernova Search Team and the Supernova Cosmology Project. The data tilt upward, indicating that the Universe is accelerating. Both plots show comparison of the 1998 Type IA supernova data to a variety of hypothetical universes with varying contributions from matter and dark energy. The bottom panel shows the difference between data and models from the $\Omega_M = 0.3$ and $\Omega_\Lambda = 0$ prediction. The best fit to the data requires

seemed ugly. Given our current knowledge of physics, dark energy doesn't make any sense. But nature sometimes surprises us and forces us to reevaluate our mindset.

In the decade following these original supernova observations, many additional measurements have weighed in on this debate. Skeptical scientists examined the dust obscuration and uniformity of light curves in great detail. Most astronomers would agree that these two issues have been settled. New Type IA supernovae have been found that corroborate the original data. An additional important test of dark energy required observers to find even older Type IA supernovae. Theorists had predicted that, prior to the current era of acceleration, there must have been an earlier epoch of deceleration. They urged the observers to confirm this turnaround.

Using the Hubble Space Telescope, Adam Riess and collaborators managed to eke out enough data to observe this transition from early deceleration to recent acceleration. They found a few ancient supernovae from 10 billion years ago (at $z = 2$). These explosions turned out to be brighter and thus closer than predicted by the ordinary expansion of Hubble's law. Their light comes to us from an earlier, decelerating phase of the Universe, prior to the acceleration occurring in the current epoch.

Eventually even the skeptics were convinced. Dark energy exists. What finally clinched it for me was the consensus with all other cosmological measurements. A variety of apparently unrelated observations all forced the same conclusion. Cosmic microwave background (CMB) data from the WMAP satellite (discussed in Chapter 3) required the existence of dark energy to explain the locations and heights of peaks in the temperature anisotropy. Quantitative measures of the observed clumping of matter into galaxies and clusters made sense only if dark energy is the major ingredient in the Universe. It's like a jigsaw puzzle with a missing piece: once a dark energy component is included,

---

the fraction of the Universe consisting of matter (both atomic and dark matter) to be $\Omega_M = 0.3$ and the fraction of the Universe consisting of dark energy to be $\Omega_\Lambda = 0.7$. Thus roughly 30% of the Universe's content is matter, and 70% is dark energy. More recently, these numbers have been modified to 31% matter (5% atomic matter and 26% dark matter) and 69% dark energy. *Adapted from Riess, A. G. 2000. "The Case for an Accelerating Universe from Supernovae."* Publications of the Astronomical Society of the Pacific *112: 1284; based on Riess, A. G., A. V. Filippenko, P. Challis, A. Clocchiatti, A. Diercks, et al. 1998. "Observational Evidence from Supernovae for an Accelerating Universe and a Cosmological Constant."* Astronomical Journal *116: 1009; and Perlmutter, S., G. Aldering, G. Goldhaber, R. A. Knop, P. Nugent, et al. 1999. "Measurements of Omega and Lambda from 42 High-Redshift Supernovae."* Astrophysical Journal *517: 565.*

everything falls into place. The dark energy has to contribute almost 70% of the content of the Universe to explain the data from these multiple experiments.

Figure 9.3 shows a plot of the dark energy fraction versus the matter (atomic plus dark matter) fraction of the Universe. The blue curve indicates the region inferred from the supernova data, the green region from the CMB, and the orange region from clusters of galaxies. All these varying types of measurements together imply a universe consisting of roughly one-third matter and the rest dark energy.

The most precise recent breakdown of the components of the Universe is 31% matter (5% atomic matter and 26% dark matter) and 69% dark energy. These numbers were obtained in March 2013 using CMB data from the Planck satellite, as discussed in Chapter 3. The atomic component is known to great accuracy to be close to 5%. The exact amounts of the two dark components, though roughly correct, are still in flux as new data come in. The most important new result of the supernova measurements is that dark energy is the predominant constituent of the current Universe.

In 2011 the Nobel Prize in Physics was awarded to Saul Perlmutter, Adam Riess, and Brian Schmidt for their discovery of the accelerated expansion of the Universe (Figure 9.4). The prize can only be awarded to a maximum of three recipients at one time. If it weren't for this restriction, two other astronomers might have shared in the prize for their seminal contributions to the discovery of dark energy: Robert Kirshner of Harvard University and Nicholas Suntzeff of Texas A&M. Suntzeff, together with Schmidt, founded the High-$z$ Supernova Search Team. Kirshner was the PhD advisor of Riess and postdoc mentor of Schmidt. It takes an outstanding team to make such measurements, and all these people played important roles.

Bob Kirshner once rescued me from a motorcycle gang. I met him when I first arrived at the Aspen Center for Physics in 1982 for a summer program in cosmology. At the time he was a professor in the Astronomy Department of the University of Michigan before leaving for the Harvard / Smithsonian Center for Astrophysics in 1990. My PhD advisor Dave Schramm had urged me to spend a month in Aspen for the science and the networking opportunities. Dave had just bought a house about a mile away from the center and said I could stay there even before he had moved in. I arrived a week before he did. From the airport I took a taxi to the center. Because Dave hadn't told me to formally apply, I was unexpected and was met by hostile stares from the secretaries. Luckily, Bob offered me a ride to Dave's house. When we got there, we found the driveway full of motorcycles, and the house occupied by enormous bearded men lounging on couches and drinking beer. The previous

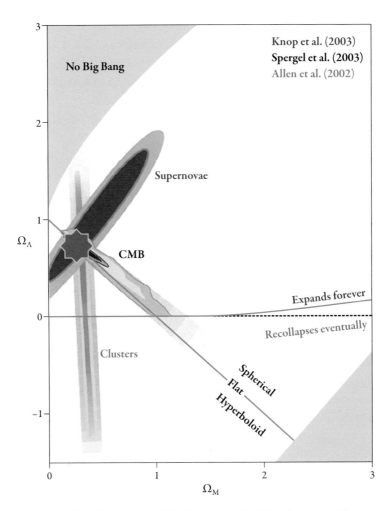

FIGURE 9.3 (A color version of this figure is included in the insert following page 82.) A variety of astrophysical data sets converge on a Universe with 31% matter and 69% dark energy. The horizontal axis is the fraction of the Universe consisting of matter (both atomic and dark matter), and the vertical axis is the fraction consisting of dark energy. In the color version of the figure, the green regions are consistent with supernova data; the red regions with cluster data; and the brown regions with cosmic microwave background (CMB) data. The best fit that matches all the data (indicated by a star) is 31% matter and 69% dark energy. Although this figure doesn't make the distinction, the matter content is further divided into 5% atomic matter and 26% dark matter.

FIGURE 9.4 Winners of the 2011 Nobel Prize in Physics: Brian Schmidt, Saul Perlmutter, and Adam Riess (left to right), "for the discovery of the accelerating expansion of the Universe through observations of distant supernovae." *Markus Marcetic.*

owners hadn't moved out yet. When Bob and I arrived, these guys promised to leave by nighttime, so that I could move in. I was terrified of spending the night there alone, when I clearly wouldn't be the only person with a key! Bob instead invited me to stay in the third bedroom in the apartment he was renting together with another astronomer, Gus Oemler.

Bob Kirshner and Gus Oemler were famous for their discovery in 1981 of "The Great Void" together with Paul Schechter and Stephen Shectman. This patch of sky is one of the emptiest places in the Universe. It is also known as the Boötes Void, because its center is located near the constellation Boötes (pronounced "boo-OH-tease"). The void is about 700 million light-years from Earth and about 250 million light-years in diameter, or roughly 2% of the diameter of the observable Universe. This region was later found to contain a few galaxies, but they are extremely sparse.

I stayed in the apartment with Bob and Gus until my PhD advisor arrived later in the week. The reason that the Aspen Center for Physics is so successful and so important is that junior and senior scientists get to know each other in an informal setting. I spent time with Bob Kirshner, Gus, and many other well-known cosmologists that summer. This type of networking is responsible for launching many successful careers as well as stimulating great ideas in all branches of physics. In addition to being a brilliant astronomer, Bob is one of the wittiest people I have ever met. His style of humor is a lot like that of David Letterman. They even look alike. Bob tells the tale of the story of the discovery of dark energy in his book *The Extravagant Universe*.

## What Causes the Acceleration?

All matter, including ordinary atoms as well as dark matter, causes gravitational attraction rather than repulsion. The atoms in the Universe together with the dark matter produced clumpiness—the planets, stars, galaxies, clusters, and other structures. Matter of any kind would act as a drag on the expansion of the Universe and cause deceleration rather than acceleration. Consequently, prior to the discovery of dark energy in 1998, cosmologists had expected to find that the Universe is decelerating.

Now that the acceleration of the Universe has been discovered, cosmologists have to postulate something completely new and exotic—a novel component with negative pressure and an effective "antigravitational" force. Michael Turner first coined the term "dark energy," and the name stuck. Cosmologists are pursuing two main theoretical avenues toward understanding dark energy: vacuum energy and modifications to Einstein's equations. I was one of the original proponents of the second approach.

The most popular, and perhaps the most likely, explanation for the dark energy is vacuum energy. Contrary to the connotation of the word "vacuum," empty space is in fact seething with energy. It is full of virtual particles and antiparticles that spring into and out of existence at every point in the Universe. As an example, an electron-positron pair can emerge from the vacuum and then immediately annihilate again. The lifetimes of these particles are so infinitesimal that we call them "virtual." As short-lived as they are, the resulting energy is large and measurable.

In 1948 the Dutch physicist Hendrik Casimir proposed an experiment to detect the effects of virtual particles. He suggested placing two uncharged metallic plates in a vacuum just microns (millionths of a meter) apart from each other. Virtual photons in the vacuum should produce an attractive force between the two plates. About 50 years later, in 1997, Steve Lamoreaux succeeded in measuring this tiny force, known as the Casimir effect. The plates

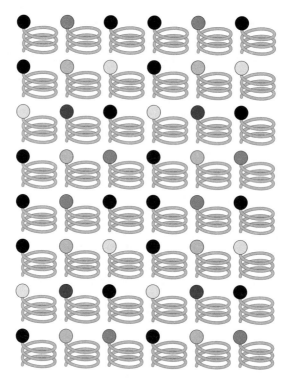

**FIGURE 9.5** Quantum field theory description of the vacuum.

inched closer together by the predicted amount. The vacuum really is full of quantum mechanical fluctuation energy. On the global scale of the Universe, this energy may be responsible for the Universe's acceleration.

Another description of the vacuum emerges in quantum field theory, a modern mathematical formalism describing particles and their interactions. In Figure 9.5, the springs with balls attached represent different particles. One ball might be an electron, while another is an up quark. Even when there are no particles present, there is still energy associated with the springs, known as ground state energy. The ground state energies of all the particles in the Universe add up to a huge amount. In fact, a naive calculation produces an unacceptably large number. The resulting energy, known as the "cosmological constant," has a value that is too large by a factor of $10^{120}$. This enormous vacuum energy would have overpowered everything else in the Universe from the beginning. It would have prevented any galaxies from forming and life from ever existing. The next section examines this cosmological constant problem in more detail.

In the 1920s, Einstein first introduced the notion that the Universe could have a large vacuum energy. He added a cosmological constant term to his equations, symbolized by the Greek letter Lambda ($\Lambda$). At the time this was just an arbitrary constant, and Einstein left its numerical value indeterminate.

A cosmological constant is unchanging in time or space. It has the weird consequence that, as the Universe expands, the total amount of vacuum energy inside an expanding region increases with time. This behavior is completely different from our ordinary experience, where matter and radiation energy densities decrease with expansion. A nice analogy is given by an automobile engine. When a piston in an automobile is pulled outward, the total number of gas atoms inside the cylinder remains constant, while the volume increases. Thus the gas density (the number of gas molecules in a hypothetical fixed volume) decreases. In contrast, if a similar experiment were possible with vacuum energy, the vacuum density would remain constant. The total amount of energy inside the cylinder would increase as you pulled out the piston.

Einstein introduced the cosmological constant in his equations of General Relativity to balance against other terms and so allow for the existence of a static Universe. Without this additional $\Lambda$ term, the solutions to the equations required the Universe to expand or contract, notions that Einstein originally found repulsive. Physicists consider symmetries elegant, and the idea that the Universe could be symmetric in time as well as homogeneous and isotropic in space seemed beautiful. Yet even theoretically, this kind of balancing act between the different terms in the equation is unstable. It is known as "fine-tuning." If there were one additional atom in the Universe, the equilibrium would fail, and the static Universe would start to move. In this sense the idea of a static Universe was aesthetically disfavored from the get-go.

As we've seen, in 1929 Edwin Hubble found that the Universe is expanding. This discovery prompted Einstein to abandon the static Universe and the cosmological constant as his "biggest blunder." At the same time the Dutch physicist Willem de Sitter found another solution to Einstein's equations. He imagined a universe containing only a cosmological constant with no matter at all. Such a universe would expand exponentially—giving rise to acceleration. Although a universe without matter could not be the one we inhabit, this idea was not forgotten. From a purely theoretical standpoint, it remained an intriguing possibility. In light of new data over the past 15 years, these old ideas have been resurrected as possible ingredients in understanding our Universe.

The discovery in 1998 that the Universe's expansion is accelerating led to a revival of Einstein's cosmological constant. Today it is considered the "vanilla" type of dark energy, the most likely of the alternatives. Yet it suffers from a deep unanswered puzzle. We've seen that in the modern quantum field–theoretic perspective, every point in the Universe is seething with vacuum energy. Particle-antiparticle pairs pop in and out of existence. The amount of vacuum energy can be quantitatively computed by adding up the ground state energies of every particle in the Universe using the spring analogy (recall Figure 9.5). The problem emerges when theorists sum up these ground state energies and find a value for the cosmological constant that disagrees with observations. As mentioned in the previous subsection, the calculation predicts a value that is too large by a factor of $10^{120}$. Scientists were already aware of the discrepancy between this gargantuan number and our Universe several decades before the supernova observations of dark energy.

This deviation by a factor of $10^{120}$ between the predicted and measured values of $\Lambda$ is known as the *cosmological constant problem*. If the value of $\Lambda$ were in fact the theoretically predicted large number, our Universe would be the one considered by de Sitter. The matter content would be essentially irrelevant compared to the big vacuum component, and the Universe would have been accelerating from the very beginning. Such a Universe would be completely unable to support life of any kind and is clearly not the world we inhabit. The cosmological constant problem is an unanswered mystery and is thought to be one of the major problems in all of fundamental physics. Its resolution may well lead to a revolution in our understanding of the Universe.

Prior to the supernova results, physicists simply set the value of the cosmological constant to zero in the equations without explanation. Because the theoretically expected value was too large by such a ridiculously huge value of $10^{120}$, the assumption was that some unknown dynamical mechanism must have driven it to zero. Many theories were proposed, but no satisfactory solution existed. Although scientists couldn't find an explanation, they took some solace in the idea that a value of zero was at least simple and therefore plausible. The theoretical bias that $\Lambda = 0$ was strong. The theoretical physics community as a whole resisted the mounting observational evidence for dark energy.

In contrast, some astrophysicists who paid close attention to the data were arguing for the existence of a cosmological constant.[6] They based their arguments on a variety of observations. Data from galaxies and clusters of different sizes pointed to the existence of a vacuum energy. In addition astronomers struggled to explain the conundrum that some stars appeared to be older than the Universe itself. A nonzero value of $\Lambda$ could resolve this discrepancy. Many of the rest of us (including me) found the idea of a cosmological constant repellant. But we were wrong. The Universe is accelerating, and dark energy exists. The simplest (though not only) explanation is vacuum energy.

## Coincidence Problem

The value of the cosmological constant $\Lambda$ consistent with the observations of Type IA supernovae created a second puzzle even more perplexing than the original cosmological constant problem. Though large enough to be the dominant component of the Universe today, the numerical value of $\Lambda$ is a tiny fraction ($10^{-120}$) of the number expected from quantum field theory. But to explain acceleration, it must be nonzero. This minuscule number is extremely problematic. Although $\Lambda = 0$ would have been hard to explain, a small value for $\Lambda$ makes even less sense.

Now two problems exist. First, there is the original cosmological constant problem. Why isn't the observed value of $\Lambda$ the huge number predicted by theory? Second, how can the vacuum energy have such a small nonzero value? Where does it come from? Explaining the origin of a small number is even more difficult than explaining a value of zero.

This second issue has come to be known as the *coincidence problem*. Because of its small value, the vacuum energy was unimportant throughout most of the history of the Universe. Matter and radiation dominated, and the Universe decelerated. However, as time went on, the matter content was diluted by the expansion, whereas the cosmological term remained constant. Eventually the dark matter lost out to the dark energy. The Universe switched to acceleration. What is bizarre, however, is the timing. The epoch when dark energy kicked in as the dominant component coincided with the epoch where the conditions became ripe for the existence of life. Cosmologists are struggling to explain this strange coincidence.

Again the analogy with an automobile engine is useful. It can illustrate the transition from a matter-dominated to a vacuum-dominated Universe. As a piston is pulled outward, the total number of gas atoms inside the cylinder remains constant while the volume increases, so that the gas density decreases. Similarly, the matter density in the Universe decreases with the expansion. In contrast, the vacuum energy density has remained constant. Eventually the decreasing matter density dropped below the value of the vacuum energy density. This switch from matter domination to dark energy domination happened roughly 6 billion years ago (at redshift $z = 0.5$, when the Universe was one and a half times as dense as it is now).

The precise value of $\Lambda$ determines the exact timing of the transition from matter to dark energy domination. Today the vacuum energy density in the

Universe is a little more than double the cosmic matter density. It is hard to imagine any aspect of fundamental physics that would drive these two quantities to comparable values at exactly the current epoch in the history of the Universe. The origin of the value of $\Lambda$ required to explain the dark energy is a complete mystery.

Steven Weinberg, professor at the University of Texas, won the Nobel Prize for his work in the 1960s on the electroweak theory—the unification of the electromagnetic and weak forces between elementary particles.[7] Among his more recent interests is the cosmological constant problem. His research took an interesting direction using *anthropic* arguments, which seek to explain properties of the Universe by arguing that they are the only ones consistent with the existence of life. Weinberg noted that the range of values of the cosmological constant that allow life to exist is narrow. If vacuum energy had dominated throughout the history of the Universe, life could never have emerged. In 1987, a decade before the discovery of dark energy, Weinberg computed the maximum value of $\Lambda$ consistent with human existence. If $\Lambda$ were larger, then the dark energy would have become important much earlier in the evolution of the Universe. Galaxy formation could never have taken place. There would be no suitable location for life forms. Weinberg argued that we should not be surprised to exist in a Universe with a small $\Lambda$, because with a larger value we could not exist at all.

Ten years after this work, dark energy was discovered. Yet the numerical value of $\Lambda$ consistent with the data is somewhat lower than Weinberg's calculation (he had predicted its most likely value to be the maximum allowed by galaxy formation). With Hugo Martel and Paul Shapiro, Weinberg has a new variant on his original ideas. These authors postulated that the Cosmos might consist of a multitude of "subuniverses," such as our own. Each has a different value of $\Lambda$ and a different epoch of galaxy formation, which again Weinberg took to be an essential criterion for the existence of life. The authors argue that the most likely value of $\Lambda$ is the one consistent with galaxy formation in a typical subuniverse—not just in ours. With this premise, the probability of living in a universe with the current observed value of $\Lambda$ becomes roughly 10%. Although this value is a little low, it is not unreasonable.

Some theorists have taken these arguments one step further. They postulate that there is a multiverse, a host of universes, each with different values of *all* the fundamental constants: different strengths of the electromagnetic force, different forces of gravity, different particle masses, and different values of the cosmological constant. These scientists suggest that we should not be surprised that we live in a world that is consistent with our own existence. They argue that every universe has a different value of the cosmological constant but most of them are uninhabitable, because their vacuum energies are too large. To clarify the argument, these researchers refer to planets as an analogy and point out that similar anthropic arguments apply. Clearly there are many different types of planets, with different masses, compositions, orbital periods, atmospheres, and so forth. We cannot expect all of them to look exactly like Earth, just because this is the only planet we can live on. Our home is not particularly typical of all planets in all solar systems. Believers in anthropic arguments point out that a similar line of reasoning could apply to the cosmological constant. It may not be reasonable to expect our Universe to be typical of all possibilities, just because it is the only one we can survive in.

One of the major motivations for taking seriously the idea of a multiverse is that it appears naturally in string theory. The goal of string theory is to provide a complete description of nature on the smallest scales and back to the earliest times. The fundamental objects are strings, and the vibrations of the strings give rise to particles. A major success of this approach is that it automatically provides a quantum theory of gravity. One single mathematical framework can describe all the particles and all the forces. However, because no testable predictions have emerged from the theory, it is not clear whether it is the correct description of our world.

String theory predicts a huge number of possible values of the vacuum energy, as many as $10^{500}$. For mathematical consistency, the theory requires the existence of 10 dimensions. In our everyday experience, we are familiar with three spatial dimensions (up/down, right/left, and forward/backward) and one time dimension. String theory predicts six additional spatial dimensions. Our three-dimensional world has a flat geometry. In contrast, these extra dimensions can have very complicated topologies and may be curled up into such tiny rings that we are not aware of their existence. An analogy is the weave in a carpet that is only noticeable if we peer very closely at it. There are roughly 10,000 possible shapes (known as Kähler manifolds) for these extra dimensions. An example is a torus, or a donut. These shapes can be threaded by a huge number of possible fluxes (similar to electromagnetic flux). As an example, one could picture French fries passing through a donut hole. Each of these possible shapes and fluxes leads to a different vacuum energy for the universe.

String theory predicts a multiverse of universes, each with its own value of $\Lambda$. Consequently some theorists believe that we no longer need to question the origin of the observed $\Lambda$. These scientists argue that we simply exist in one of the many possible universes. Clearly we must live in one that is capable of harboring life. This multiverse perspective is strengthened by theories of inflationary cosmology, an early accelerating period of the history of the Universe. Here, again, there may be multiple possible universes with different vacuum energies. The situation is quite perplexing.

A similar argument applies to the strength of electromagnetism. If the force between electrons and protons were any different, then nuclei would be unstable. We could not exist. Why does this force have exactly the right value? Again, some would say that there are many universes with many values of the electromagnetic force. The strength of electromagnetism is determined by the size of an extra dimension and could in principle have any value. These scientists would argue that it is not a mystery why humans live in one of the universes that allow for stable nuclei.

I am not a fan of the multiverse. I would prefer to have a dynamical argument for the origin of each of the properties of our Universe. I think we have to explain the value of $\Lambda$. We have to explain the strength of electromagnetic forces. It is not enough to say these have the only values consistent with human life. It is certainly plausible that there are many universes out there. Yet saying that we happen to live in one universe of many doesn't absolve us of explaining why our world behaves the way that it does. I don't think we are excused from solving the physics problem.

This disagreement is currently a big one in the cosmology community. Some prominent scientists believe in the multiverse. They are trying to quantify how many of the possible universes would have properties similar to ours. The rest of us are continuing to work on explaining as much as we can in the context of the one Universe we live in.

### Time-Dependent Vacuum Energy

Another explanation for the acceleration of the Universe is a vacuum energy that changes with time. Unlike the cosmological constant $\Lambda$, which always retains the same (constant) value throughout the history of the Universe, a different type of vacuum energy could have started out with a large value in the early Universe and decreased as time went on.

I first wrote a paper on a decaying vacuum energy in 1988, a decade before there was any experimental motivation for it. My collaborators and I were interested in observational consequences of a proposed solution to the cosmological constant problem. Emil Mottola and Pawel Mazur had suggested that mass and energy could cause a gravitational backreaction on $\Lambda$ that would cause it to drop from $10^{120}$ to a more reasonable value. Others had previously proposed similar ideas, but the technical aspects of the calculation were tricky.

Unfortunately this particular approach, although interesting, is extremely difficult to implement, and it is not clear whether it works.

Motivated by these ideas, we worked out the cosmological consequences of decaying vacuum energy. My collaborators on the project were my housemate Emil Mottola (the originator of the motivating model), my former fiancé Josh Frieman, and my then-fiancé Fred Adams. It was quite a collaboration. We called the project "rho vacuum," where the word "rho" spells out the Greek symbol $\rho$ (standing for energy density). The four of us met regularly in my (not so large) office on the sixth floor at the Institute for Theoretical Physics, situated on the campus of the University of California, Santa Barbara. Josh would open the window because he thought it was stuffy. I would close it again because I was cold. Emil would stubbornly persist in results that the rest of us thought were wrong. By 8 p.m. I would be starving and would insist on heading over to Isla Vista, the part of town where the students lived, to grab Chinese food. Emil would get exasperated and demand that I wait another half hour (which of course would turn into 2 hours).

Fred invented $w$ as the ratio of the nonvacuum pressure to the total density. I couldn't understand why he cared about that quantity. That definition turned out to be quite prescient. Nowadays $w$ is instead taken to be the ratio of the vacuum pressure to the density, and everybody is trying to measure it. For the case of a cosmological constant–dominated Universe, the value is $w = -1$. Other explanations for an acceleration can be treated as having an effective $w$ with different values. As long as $w < -1/3$, the Universe is accelerating. In fact, it was my irritation with this obsession with $w$ that caused me to come up with an alternative that I describe below.

A year after our paper, Jim Peebles, Bharat Ratra, and independently, Christof Wetterich explored similar ideas but with a different motivation. A major difference between our work and theirs was that they introduced a scalar field (similar to the Higgs field of Chapter 6) as the origin of the time-dependent vacuum, whereas we just took it to vary freely with time. Such scalar fields are thought to be the origin of an early accelerated phase called inflation, just fractions of a second after the Big Bang. These scientists proposed a similar mechanism for a rolling scalar field producing a time-dependent vacuum in the Universe's recent past.

Ten years later, their model offered an interesting explanation for the acceleration of the Universe. Paul Steinhardt, then at the University of Pennsylvania, substantially elaborated on these ideas. He produced a variety of creative models that could explain the supernova data. He named the basic approach of a rolling scalar field for the dark energy "quintessence." Time-changing vacuum energy became one of the major directions of theoretical research on dark energy. Unfortunately, the coincidence problem is not resolved in any of these models. The scalar still has to roll down to precisely the right value of the vacuum energy today. As in the case of a cosmological constant, there is no explanation for the coincidence of the timing.

### Modifying Einstein's Equations as an Explanation of Dark Energy

Cosmologists are also pursuing a second theoretical avenue toward understanding dark energy: modifications to Einstein's equations. In lieu of a vacuum energy, it is possible that the equations we use to describe the expansion of the Universe are incomplete. Although they beautifully describe everything else about the evolution of the Universe, perhaps on the largest scales they need some tweaking.

In 2002 I was sitting at a conference in Chicago on dark energy. The focus of the meeting was to propose observational tests of $w$, the parameter (discussed in the previous subsection) characterizing the amount of negative pressure required to explain the Universe's acceleration. The description in terms of a single parameter seemed to me to be too primitive. Out in the hallway, I had a conversation with two colleagues who shared my reservations. Suddenly I had a flash of insight. It occurred to me that I could apply some of my recent work on extra dimensions to explain dark energy.

With postdoctoral fellow Daniel Chung (now professor at the University of Wisconsin), I had been thinking about braneworlds. Motivated by string theory, theorists were treating our three-dimensional Universe as a membrane (or "brane"), much like the surface of a drum, situated in additional spatial dimensions. Most Standard Model particles are confined to remain on the brane. Only gravitons (which are responsible for gravitational forces) and new nonstandard physics can exist in the extra dimensions. Dan and I had realized that the contents of the extra dimensions could pull on our three-dimensional brane and produce modifications to the equation for the Universe's expansion. We had gotten into arguments with others who insisted that only one specific variant on the equation was possible. We realized that in fact the modifications could be just about anything.

Out in the hallway it occurred to me that I could use these altered equations to explain accelerated expansion without resorting to any contents in our Universe other than matter and radiation. No vacuum was required! When I got back home, my graduate student Matthew Lewis and I wrote a paper in a week and submitted it to the journal *Physics Letters*.[8] It was accepted immediately. Our original paper focused on a specific simple addition to the terms in the equation for the Universe's expansion but could easily be generalized to alternate variants.

I guess we made a mistake in the name we gave the model. I had heard Lisa Randall of Harvard University talk about "the warp factor" in her well-known work with Raman Sundrum.[9] They used warped extra dimensions as a solution to the hierarchy problem in particle physics (the mass difference between real-world particles and the enormous Planck mass of $10^{19}$ GeV). Since I'd always heard the warp factor called the conformal factor instead, I assumed she had gotten the name from *Star Trek*. Instead it turns out it's a relativity term from decades ago. Based on my misunderstanding of the origin of the terminology, I got the brilliant idea to look to *Star Trek* for a name for our new model of the accelerating Universe. I came up with "Cardassian expansion." I got a big kick out of it when one of my colleagues asked me, "Can you remind me, who is Cardassius?" No, there was no such Roman from antiquity. Instead, the name was based on the alien race indigenous to the *Star Trek: The Next Generation* television show. Cardassians look foreign to us yet consist entirely of matter and are bent on the accelerated expansion of their evil empire. Similarly in our new dark energy model, an accelerating universe looks alien, but it consists merely of matter without resorting to any vacuum energy. Not everyone has my sense of humor, and I think some took umbrage at the name. It certainly has not stuck.

After our paper on Cardassian expansion was published, we learned of similar work that had preceded ours. A year earlier, Cedric Deffayet and collaborators had proposed a different model based on modified gravity in extra dimensions.[10] I was disappointed to learn that our paper was one of the earliest but not the first. They too had suggested alterations in the equations describing the Universe's expansion as the origin of the accelerating Universe. The specifics of their work were different from ours, but the idea of modifying the equations was similar. Since that time, many have studied the observational consequences of modifications to Einstein's equations. Paolo Gondolo and I started a program for doing this but then moved on to other research topics. Others have devised a systematic approach to look for tests of modified gravity as a mechanism for explaining accelerated expansion.

After graduate school Matt Lewis, my collaborator on Cardassian expansion, took his expertise in cosmology and applied it to interesting work at General Dynamics Corporation. He even used quantum field theory in curved spacetime to help missiles hit their targets.[11] Matt has won many awards for his work and is now a director at Michigan Aerospace.

In all these alternate gravity models, including Cardassian expansion, the effective Einstein's equations describing our Universe's expansion are modified from the standard Hubble law. In lieu of a vacuum energy, these altered equations play the role of dark energy. In the early Universe, these modifications are unimportant. Only recently in the history of the Universe, approximately 8 billion years ago, did the new term in the equation kick in and start to dominate the expansion history. At that point, even though the only contents of the Universe are matter and radiation, the Universe started to accelerate.

All modified gravity models suffer from the coincidence problem. In the case of a model with a cosmological constant, this problem can be phrased in terms of the value of the vacuum energy: Why does the vacuum start to dominate around the time when life could start to exist? In modified gravity models, a similar problem emerges: the new term in the equations becomes important just as life begins to form. In both cases, the timing is a coincidence that has to be put into the theory by hand.

On the observational side there is currently a big push to get more information about dark energy. Astrophysicists want to test its pressure and its time dependence and to look for evidence of possible modifications to Einstein's equations. The approaches are varied, from studies of the evolution of galaxy clusters, to baryon acoustic oscillations (the matter equivalent of the ripples in the CMB), to finding more Type IA supernovae in both the near and distant universe. In August 2013 the Dark Energy Survey started taking data that will map one-eighth of the sky. Located in the Andes Mountains in Chile, the telescope will observe thousands of supernovae as well as hundreds of millions of galaxies to learn about the nature of dark energy.[12]

Unfortunately, infrared light from the most distant supernovae is obscured by atmospheric dust. For this reason telescopes here on Earth cannot observe them, and a new space satellite will be required. Such a mission costs billions of dollars. Although a proposed satellite named Joint Dark Energy Mission had been ranked by the Department of Energy as one of its top particle-physics priorities, political pressure killed the project. The Joint Dark Energy Mission would have been a joint project with NASA. Ten years of exploratory research had already been funded at several institutions around the United States, including the University of California, Berkeley, and the University of Michigan. It seems likely that the European Space Agency will instead fly a satellite named Euclid to study dark energy. U.S. scientists are preparing to participate in the European project.

As in so many other experimental efforts, the United States is ceding ground to other countries. It is already true that experimental particle physics at the highest energies no longer takes place in the United States. Instead it is happening at CERN in Europe. Let's hope that astrophysics doesn't suffer the same fate. Studies show that half of U.S. economic activity is based on scientific advances, so this trend is frightening.

Recently the Department of Defense donated three telescopes that it no longer needs to the cosmology community. At the time the Hubble Space Telescope was built, these other telescopes were also commissioned for the military. Obtaining them is a coup for the field and could allow dark energy studies to proceed.

Cosmologists are eager to know which, if any, of the ideas for the dark energy is right. So far the vanilla model of a dark energy, consisting of a cosmological constant, is winning. Every piece of data obtained to date is consistent with a simple vacuum energy. The uncertainties are getting smaller, and no deviation has yet been found. One of the major observational goals is to distinguish a constant from a time-dependent vacuum.[13] This involves a host of difficult measurements and will take a long time. In the meantime theorists are continuing to try to make sense of this mystery given the information that we currently have.

## The Future of the Universe

The acceleration of the Universe has forced cosmologists not only to revise their understanding of the past and present evolution of the Universe but also to revisit predictions about its future. Speculation about the future evolution of the Universe requires a great deal of hubris. How can we possibly think we understand everything well enough to answer such questions? Yet, given what we do know, we can try.

### Shrinking Horizon

The existence of dark energy may have dire consequences for the future of the Universe. One unfortunate corollary is that we had better build as many observatories as we can now. Because all stars are accelerating away from us, we can learn more about them now than we will be able to in the future—they are moving away so fast that in the future, their light will not be able to get to us!

Telescopes are time machines. Because it takes time for the light from distant objects to reach us, in fact we are seeing them as they were in the past, not as they are now. As mentioned at the beginning of the chapter, the farther out we look, the farther back in time we can study the Universe. Light travels at a velocity of $3 \times 10^8$ meters per second, and according to Einstein's Special Relativity, nothing can travel any faster. When we look at the Sun, we see it as it actually was 8 minutes ago, because that's how long it took the light to reach us. If the Sun blinked out of existence, we would be blissfully ignorant for 8 minutes before the world went dark and our surroundings started to get colder and colder. First the atmosphere would cool off, and gradually the core of Earth itself would cool. In a different doomsday scenario, a nearby star could turn into a supernova that engulfs the entire Solar System, and we wouldn't know until it was too late. By the time the neutrinos and the light from the supernova reached us, we would survive only another few minutes before being killed by the oncoming shock waves.

Thanks to ever-more-powerful telescopes, astronomers can look out into the more distant Cosmos and consequently farther back in time. However, there is an absolute cutoff set by the distance that light has been able to travel since the Big Bang. This distance, known as the horizon length, is determined (by multiplying the speed of light times the age of the Universe) to be 13.8 billion light-years, or equivalently, $10^{26}$ meters. This length scale sets the size of our observable universe: anything farther away is completely inaccessible to us.

Throughout most of the history of the Universe since the Big Bang, the horizon length became larger. As time went on, light had more time to travel longer distances and convey information from one point to another. The size of the observable Universe grew during the early decelerating phases of the Universe. Until recently, cosmologists assumed that the horizon size would continue to grow as the Universe aged. However, the acceleration of the Universe has forced us to rethink this conclusion.

Once the Universe started to accelerate roughly 8 billion years ago, the horizon started to shrink. Galaxies and other structures accelerated away from one another (because of the expansion of space) faster than light rays could carry information between them. As a consequence, the fraction of the Universe that we can access is now decreasing with time. Sadly, any astronomical objects that we do not study now will forever move outside our realm of accessibility as time goes on. We could joke that NASA should fund as much space science as possible now! Of course, if humans are able to harness wormholes or other mechanisms to beat the speed of light, then we may be able to travel to far distant places in regions of the Universe currently outside our grasp.

### The Big Chill

The final fate of the Universe billions and trillions of years from now depends on its geometry. Of the three possibilities posited in 1930, only one had the capability of recollapsing: the spherical geometry (see Chapter 3). The amount of mass and energy in this case would have been so large that it provided enough gravitational pull to cause the sphere to reach maximum expansion and then start recollapsing. Eventually, the sphere would collapse to a high-density Big Crunch. Either this would be the end of the Universe, or it would rebound into an oscillating universe with alternating periods of expansion and collapse. The other two geometries, flat or hyperboloid (saddle-shaped), would contain less mass and energy than the spherical case. Both of these would expand forever, becoming less and less dense and colder and colder. The end state for these two cases would be a Big Chill.

Because CMB measurements reveal that the Universe has a flat geometry (as discussed in Chapter 3), it seems likely that we are headed for a Big Chill. It remains possible that our Universe is a giant sphere, so large that its curvature is imperceptible to CMB measurements. In that case it will eventually recollapse after an enormously long period of expansion. However, the most likely scenario for the future is a Big Chill. As the Universe continues its accelerated expansion, matter will become more and more diffuse. Galaxies and other structures will be ripped apart to leave only a bleak empty Universe.[14]

### The Ultimate Fate of Life

An even more disconcerting consequence of dark energy is the dismal fate of life in the Cosmos. As the Universe hurtles to a future where all matter becomes more and more spread out, the world will become darker and darker and colder and colder. Is there hope for life in such a Universe—certainly not life as we know it, but any form of life at all? The answer depends on the detailed nature of dark energy.

In 1979 Freeman Dyson first attempted to quantify scientifically the question of the ultimate fate of life in an expanding universe.[15] He proposed a framework in which to discuss whether some form of life, material or otherwise, can go on. Although human life cannot persist in an eternally cooling universe, still some other type of life performing computations could in principle persist indefinitely. Dyson's condition can be stated in the following way: life can be considered "infinite" if the number of computations, or "subjective time," can be infinite while the total energy consumed is finite. It at first appears possible to achieve this criterion as long as the operating temperature of the organism continuously decreases in time. However, the creature must be able to get rid of the heat generated by the computations it performs. Counterintuitively, even though the Universe is perpetually cooling, the creature will fry to death unless it can dissipate the heat that it creates. Dyson estimated an upper limit to the rate at which waste heat can be radiated and found scenarios where indeed life could go on indefinitely. It would need to hibernate periodically to achieve this goal, but it could be done.

However, at the time of Dyson's work the Universe was assumed to be decelerating. In the light of evidence that the Universe is accelerating, the conclusions of his original work required reinvestigation. Lawrence Krauss and Glenn Starkman followed Dyson's basic approach to reexamine the future in a Universe dominated by a cosmological constant. These cosmologists concluded that life is inevitably doomed to oblivion if the value of the dark energy density remains constant.[16] Any life form would eventually fry to death in the bath of thermal Hawking radiation produced by the vacuum energy—much like the Hawking radiation from a black hole that was introduced in Chapter 5. Beings of any kind generate heat by the process of living and would eventually be unable to dissipate their heat in the background of this thermal bath.

William Kinney of the University of Buffalo and I showed that this heat death could be avoided if the dark energy density is not a constant but instead decreases as time goes on. We studied the consequences of other explanations for the acceleration of our universe: a decaying vacuum energy, quintessence, or Cardassian expansion. In these models the temperature of the cosmological Hawking radiation decreases in time, in many cases quickly enough to allow life to continue indefinitely despite the presence of the thermal bath. We argued that if the acceleration is caused by anything other than an unchanging cosmological constant, the time dependence allows life to escape the ultimate heat death.

In *New Scientist* (the British equivalent of *Scientific American*), Philip Ball wrote, "Katherine Freese and William Kinney may not look much like superheroes, but this pair of astrophysicists may just have rescued all life in the universe."[17] My question to him is: "How do you know we don't look like superheroes? You have never met us!"

## Epicycles

The greatest fear of cosmologists is that there is a major ingredient lacking in our current understanding of the Universe. Is it possible that dark matter and dark energy don't exist? Could scientists be missing something fundamental? Perhaps an entirely different way of looking at the world will replace the need for these invisible pieces of the Universe.

Certainly such errors have been made before in the history of science. The Greek philosopher Aristotle in roughly 350 B.C. proposed that the Cosmos is composed of crystalline spheres. He believed that all celestial objects were attached to these spheres. Earth was thought to be at the center of this system and at the center of the Universe. Circa 150 A.D., Ptolemy (who lived in Alexandria, Egypt) refined these ideas. He had the planets orbiting in small circles called epicycles; the centers of the epicycles in turn revolved in larger circles. As more astronomical data came in, the model became more and more convoluted. The system began to require epicycles on top of epicycles. Ptolemy's theories dominated astronomy for more than 1,000 years, and in the Middle Ages became part of Christian dogma. The Catholic Church believed that the central position of the Earth reflected God's interest in mankind.

In 1543 Nicolaus Copernicus proposed a radical change. He abandoned epicycles in favor of a heliocentric view of the Universe, in which the Sun is at the center of the Solar System. Although initially banned by the Church, this view eventually won, because it is correct.

In the modern era, the fear of modern astronomers is that they are again constructing the equivalent of epicycles on epicycles. Cosmologists feel certain that they understand the basic picture of the Cosmos. Yet to accommodate new data, they keep having to add new pieces. First there was the dark matter in galaxies. Now there also appears to be dark energy. Do these unidentified components really exist? Or does modern cosmology need a change in perspective that is the equivalent of moving from a geocentric to a heliocentric worldview? The disquieting alternative is that a paradigm shift is required to make sense of the data.

Yet the case for dark matter is so strong, so consistent, and so easy to resolve with a new fundamental particle. Dark energy is a little more disturbing, because scientists really don't know how to begin to explain it. Most cosmologists feel that the dark matter problem seems sensible and soluble and that experimental tests should be able to resolve it soon.


## Epilogue

The past century has seen a revolution in our understanding of our Universe. Whereas astronomers at the time of Einstein in the early twentieth century thought that the entire Universe consisted of a single galaxy—the Milky Way—cosmologists today routinely study billions of galaxies in our observable Universe and ponder the nature of an infinite Universe. At the turn of the millennium, cosmic microwave background observations in Chile and at the South Pole measured the geometry of the Universe and found it to be flat. This remarkable discovery amounted to weighing the Universe and measuring its total mass and energy density. Yet, despite this remarkable progress, the content of the Universe is basically unknown. Mysterious dark matter and dark energy pervade the Universe.

FIGURE 9.6 The panel on "The Dark Side of the Universe" at the World Science Festival in New York, June 2011. The three women representing dark matter are, from the right, the author, Elena Aprile (Professor of Physics at Columbia University), and Glennys Farrar (Professor of Physics at New York University). Then continuing to the left are three men representing dark energy: Michael Turner (Professor of Physics at the University of Chicago), Saul Perlmutter (Lawrence Berkeley Laboratory) and Brian Greene (Professor of Physics at Columbia University and co-host of the Festival). © *2011 World Science Festival.*

Any time now, the question of the dark matter may be answered—by experiments in atom smashers, space, underground mines, or phototubes at the South Pole. I am personally excited because experiments using approaches I proposed 20 years ago appear to be on the verge of detecting dark matter particles. Already some groups have unexplained results that may herald discovery.

The World Science Festival takes place in New York City every year for a week in June. Inaugurated in 2008, this wonderful event is the brainchild of Brian Greene, one of the best science writers today, together with television producer Tracy Day, the Executive Director of the Festival. Brian is a Professor of Physics at Columbia University and works on string theory. His books include *The Elegant Universe,* on the subject of string theory, and *The Hidden Reality,* on parallel universes. With an annual budget of 5 million dollars, the World Science Festival hosts public lectures and panels covering all branches of science "with the purpose of engaging the public in cutting-edge ideas and research discoveries." To quote the Festival's webpage, "In 2011 almost 200,000 people attended the Festival's events and programs, with lines wrap-

**FIGURE 9.7** (A color version of this figure is included in the insert following page 82.) "Dark matter is attractive, while dark energy is repulsive!" © *2011 World Science Festival.*

ping around a city block for possible stand-by tickets to sold-out programs." I participated in two panels at the festival in 2011, "The Dark Side of the Universe" (Figure 9.6) and "World Science Festival Salon: The Mystery of Dark Matter."

"The Dark Side" discussion had six panelists: I was one of three women representing dark matter, and three men represented dark energy. This panel was so much fun. Our moderator John Hockenberry, an award-winning journalist and radio and television host, kept a lively discussion going. Figure 9.7 shows me making a joke. Because the three women were talking about dark matter while the three men were talking about dark energy, I pointed at the men and said, "Dark matter is attractive while dark energy is repulsive." In fact, from the physics point of view, this is true!

John's last question to all of us on the panel was about our expectations for the future. All of us agreed that we expect the dark matter particle to be discovered soon, most likely within the next decade. Elena Aprile even expressed her ambitions for a more rapid discovery over the next few years with her XENON detector. The dark energy side is much more difficult, because we don't even know what we're looking for. I look forward to the solution of the dark matter problem, one of the deepest mysteries in all of modern science!

# AFTERWORD:
## DARK STARS

I n the beginning the Universe was dark, with no stars or other source of light. Then 200 million years after the Big Bang, the first stars began to form and one by one lit up the Universe. Eventually these stars burned out and spewed into the Universe the elements that are required for the beginnings of life. Now there is a new twist on this story. In 2007 my collaborators Paolo Gondolo and Douglas Spolyar and I proposed the idea for a new type of star, powered by dark matter particles, the elusive subatomic constituents that have been the subject of this book.[1]

We named these objects "dark stars" after a Crosby, Stills, Nash, and Young song of the same name. These new stars were radically different from any stars forming today. The new twist is that dark stars were powered by dark matter inside them. The energy source for today's stars, like the Sun, is the fusion of light elements into heavier ones inside their cores: hydrogen burns to helium and in the process provides the pressure to counterbalance the gravity that would otherwise cause the Sun to collapse. In contrast, in the early Universe, an alternate, more powerful heat source for stars existed: dark matter annihilation. This is the same annihilation process that leads to indirect detection searches for the photons, electron-positron pairs, and neutrinos resulting from the annihilation.

Dark stars formed at the centers of dark matter minihalos, the large spherical precursors to today's galaxies. Although there isn't enough dark matter in today's stars to produce a significant amount of energy, the first stars existed in a much richer dark matter environment. The early Universe was more dense and compact, with more dark matter concentrated in the early protostellar gas. Just as when fusion was first proposed as a power source for stars in 1920, dark matter annihilation may revolutionize our understanding of how stars work.

At the time the Universe was 200 million years old, the conditions became ripe for the formation of stars. Galaxies form in a hierarchical fashion, starting from Earth-sized spheres of dark matter. These then coalesced into larger objects, merging together into ever-larger halos, until eventually galaxies formed. Along with dark matter, atomic matter was pulled into the centers of these halos. Midway in this process of galaxy formation, halos of 1 million solar masses formed, and it was inside these halos that protostellar clouds of gas could finally begin to collapse, smack in the centers of these halos, where there is a huge abundance of dark matter. Until 2007 no one had asked the question of what role the dark matter plays as these clouds collapse. My collaborators and I asked what the consequences would be of WIMP dark matter inside the protostellar clouds. When we addressed this issue, we were amazed by the results we found. If three conditions are met, the collapsing cloud turns into a dark star: there must be enough dark matter fuel, the dark matter annihilation products must not be able to escape the star, and the dark matter heating must beat all cooling mechanisms. The products of the annihilation process are high-energy particles that get stuck inside the star and heat it up. Because of this heat source, the gaseous cloud is unable to collapse any further and instead becomes a stable star—a dark star.

At first my colleagues and I didn't realize that dark stars really did shine like other stars. It took us another year to understand that dark stars, though powered by dark matter, can be incredibly bright, the brightest objects in the sky for tens of millions of years. They can be a billion times as luminous as the Sun. As a consequence, they are detectable. They may already be visible in current Hubble Space Telescope data, but the telescope's resolution isn't good enough to differentiate such stars from galaxies. The next generation instrument, the James Webb Space Telescope, will start taking data in 2018 and should be able to see dark stars. If we are right, then this telescope will discover an entirely new type of star, cool like the Sun but up to a billion times as bright.

At first dark stars are not particularly massive, weighing about as much as the Sun. Yet they are giant puffy objects; their physical size is huge, equivalent to the distance from the Sun to the orbit of Earth. By stellar standards their temperatures are cool—not hot enough for any fusion to be taking place inside the star. As long as they are cool, there is nothing to prevent more mass from falling onto them. Unlike ordinary stars, their accretion of additional material is not self-limited. Because their surface temperatures are so low, there is no ionizing radiation to prevent further accretion. More and more hydrogen falls onto them, and they grow to be huge, 1,000 to 10 million times as massive as the Sun. The existence of such supermassive stars was suggested in the 1960s, but there was no mechanism to make them until now.

Dark stars live for millions to billions of years, as long as they are fed dark matter fuel. Some may survive until today, but most of them died before now. Once they run out of fuel, they become smaller and hotter, and a period of fusion sets in. Like today's stars, they burn nuclear material. The hydrogen is converted to more complex elements that constitute the world as we know it today: carbon, oxygen, nitrogen, and so forth. Then some of this material is blown off into space, and scatters into the Universe the ingredients necessary for life.

After this brief fusion phase, most of the stellar material of the dark star collapses into a giant black hole. This emergence of black holes from dark stars may solve what has come to be known as the big black hole problem. Enormous black holes weighing 1,000 to 10 billion times as much as the Sun are ubiquitous in the Universe, and there is no good explanation of where they came from. We've seen earlier in the book that the center of every galaxy, including our own Milky Way, contains a giant black hole. Even harder to explain are billion-solar-mass black holes observed to exist already at $z = 6$, almost 13 billion years ago. The question of the origin of all these large black holes has so far eluded an answer. Dark stars may solve this big black hole problem. After running out of fuel, they collapse into black holes of 1,000 to 10 million solar masses that can serve as seeds for the giant black holes seen in the Universe.

Dark stars are remarkable beasts. One key property is that they are composed almost entirely of hydrogen and helium from the Big Bang—and just a smattering of dark matter at the level of a fraction of a percent of the mass of the star. They are enormous, with radii stretching out the distance from the Sun all the way out to Earth. Yet this small amount of dark matter suffices to power the stars: this is the "power of darkness." Particle annihilation is the strongest imaginable source of power known to science. Unlike fusion, which is only 1% efficient (at extracting power out of fusing nuclei), annihilation is almost 100% efficient: most of the mass of the annihilating particles is converted to useful fuel. Perhaps someday we will find a way to make use in our own lives of this potent material.

In addition to the three-pronged approach to dark matter detection described in this book, dark stars may provide a fourth approach. When (and if) dark stars are detected in the James Webb Space Telescope, we can learn about the properties of dark matter particles. In an ideal world, perhaps dark stars will provide definitive discovery of dark matter particles. I spent my sabbatical year in 2007 as a Miller Professor at the University of California, Berkeley, writing only one paper: the one creating this new idea of dark stars. Perhaps I will be persuaded to write a sequel to *The Cosmic Cocktail* on the subject of these dark stars.

## ACKNOWLEDGMENTS

thank many friends and colleagues. First and foremost, I am grateful to the wonderful science writer Tom Siegfried, who critiqued several chapters of the book and helped me hone my style. His advice was invaluable. I thank Debbie Liehs, who proposed the title of the book. I really appreciate the help of Bob Cousins, former spokesperson of the CMS experiment at CERN, who critiqued the section about the Large Hadron Collider in Chapter 6. I thank Dragan Huterer and Greg Tarle, both members of the Dark Energy Survey collaboration, for discussions about supernovae. My graduate student Alejandro Lopez used the lensing iPhone app of Eli Rykoff (now at SLAC National Accelerator Laboratory) to make the photo of lensed University of Michigan students in Figure 2.9. Susan Land, Juan Miguel, Aaron Pierce, Pearl Sandick, and Bob Watts read portions of the draft and gave me useful comments. I thank Andrzej Drukier, Rocco Samuele, and Gwen Tessier for their feedback on the entire book. I am grateful for the writing instruction I received in eighth grade from my teacher Mrs. Selig, throughout the years from Susan Land, and at the Harvard Extension School from Robie Macauley. I thank Isabelle Anderson for helping me to find my voice to tell my tale.

# NOTES

## Preface

1. The atomic component is known to great accuracy. The amounts of the other two components, though roughly correct, are still in flux as new data come in. The numbers listed here are the results from the Planck satellite announced in March 2013.

## ONE The Golden Era of Particle Cosmology, or How I Joined the Chicago Mafia

1. Taylor, E. F., and J. A. Wheeler. *Spacetime Physics*. New York: W. H. Freeman and Company, second edition, 1992.

2. For Christmas my son gave me a T-shirt that I love. It has a picture of Aspen mountain and it says, "Black diamonds are a girl's best friend."

3. Moguls are a series of bumps on a ski trail. These form naturally as skiers shove the ski during turns and are considered difficult terrain for most skiers. I love skiing, and fast turns on moguls are my forte. Dave Schramm had a unique style, and he simply plowed through them.

4. Weinberg, S. *Gravitation and Cosmology: Principles and Applications of the General Theory of Relativity*. Hoboken, New Jersey: John Wiley & Sons, 1972.

5. CERN is the highest-energy particle accelerator in the world.

## TWO How Do Cosmologists Know Dark Matter Exists?

1. Zwicky's collaborator on the subject of neutron stars was Walter Baade: Baade, W., and F. Zwicky. 1934. "Remarks on Super-Novae and Cosmic Rays." *Physical Review* 46: 76.

2. The only astronomical objects more compact than neutron stars are black holes.

3. As discussed further in Chapter 5, Andrea Ghez and Reinhard Genzel tracked stars moving around the center of the Galaxy and demonstrated that it hosts a supermassive black hole.

4. Though some black holes are candidates for the dark matter in galaxies, the supermassive black holes in galactic centers do not contain enough mass to account for all dark matter in the galaxy. Scientists and amateurs alike are fascinated by black holes. Chapter 5 devotes much discussion to these amazing objects.

5. Babcock, H. 1939. "The Rotation of the Andromeda Nebula." *Lick Observatory Bulletin* 498.

6. Rubin, V., and W. K. Ford. 1970. "Rotation of the Andromeda Nebula from a Spectroscopic Survey of Emission Regions." *Astrophysical Journal* 159: 379.

7. Freeman, K., and G. McNamara. 2006. *In Search of Dark Matter.* New York: Praxis Publishing.

8. Simple mathematical arguments provide numerical values of the amount of dark matter. Just as for the rotation curve of the Solar System, one can use Newton's law of gravitation to infer the amount of mass in the galaxy. In the case of the Solar System, most of the mass is concentrated in the Sun at the center. Here we have to generalize the previous formula, because the galactic mass is far more spread out. With the new formula, we can use the measured velocities of objects orbiting the galaxy to deduce the amount of dark matter. The new form of Newton's law becomes $GM_r m/r^2 = mv^2/r$. Here, $m$ and $v$ are the mass and speed of the orbiting gas, respectively; $r$ is the distance from the Galactic Center; $G$ is Newton's constant; and $M_r$ is the total mass interior to the radius $r$. The left-hand side of the equation is the Newtonian force of gravity; the right-hand side is the mass times the centripetal acceleration $v^2/r$. For the case of a single central mass like the Sun, this equation reproduces exactly the relation previously discussed for the speeds of the planets. By measuring the speed $v$ of some object at distance $r$, it is possible to calculate what $M_r$ must be (that is, the amount of mass interior to radius $r$). From rotation curves like Figure 2.6, astronomers are forced to conclude that dark matter dominates the mass of galaxies.

9. The launch of JWST has been repeatedly delayed because of cost overruns. Many cosmologists are eagerly anticipating JWST data and hope the current launch date will not shift yet again.

10. Light emitted in visible wavelengths at early times has been stretched due to the expansion of the Universe to much longer infrared wavelengths by the time it reaches Earth. Unlike HST, the upcoming JWST will be able to observe the Universe in infrared wavelengths, allowing it to see very faint objects out to great distances.

11. As we'll see in Chapter 5, dark matter particles do not have electric charge, and they do not feel the so-called "strong force" between nucleons in atoms.

12. The process of structure formation originated during the inflationary epoch, a period of rapid (superluminal) expansion in the earliest moments of the Universe. Tiny fluctuations developed: parts of the Universe had slightly more mass in them than their neighbors had, while other patches contained less mass than average. The details of these fluctuations are beyond the scope of the current discussion, but the mass differences in different regions had very important consequences.

13. We've seen above that photons were freed from atomic matter 380,000 years after the Big Bang. Those photons became the cosmic microwave background that is discussed in detail in Chapter 3. The end of the "dark ages" refers instead to the production of visible light, which awaited the creation of stars roughly 200 million years after the Big Bang.

14. The original idea for dark stars came out of collaboration with Douglas Spolyar and Paolo Gondolo.

15. A movie of this time sequence can be found at http://cosmicweb.uchicago.edu/images/mov/bnr_half4.mpg.

**THREE** The Big Picture of the Universe

1. Wendy Freedman, leading the Key Project of the Hubble Space Telescope and using Spitzer Space Telescope data for calibration, finds a best estimate for the Hubble constant of roughly 74 kilometers/second/megaparsec. However, the European Space Agency's Planck satellite finds a lower value of 67 kilometers/second/megaparsec. The two approaches are completely different, and the discrepancy is not yet understood.

2. More accurate measurements of the age of the Universe came from CMB observations described below.

3. The (near) flatness of the Universe is unexplained in the standard Hot Big Bang. In 1981 Alan Guth proposed a solution to this conundrum. He realized that an early period of exponential expansion, which he called inflation, would drive the Universe toward flatness.

4. Kelvin is a temperature scale designed so that 0 Kelvin corresponds to "absolute zero," at which all thermal movement stops. To compare to more familiar temperature scales, water freezes at 273 Kelvins, 0 degrees Celsius, and 32 degrees Fahrenheit.

5. Microwave ovens work by exciting rotations of water molecules in food. The rotating molecules collide with adjacent molecules, and the resulting energy heats the food.

6. In both the Steady State and Cold Big Bang models, there never was an epoch of primordial nucleosynthesis, the subject of the next chapter. Proponents of these alternative models had to work hard to produce exactly the right element abundances in stars.

7. Specifically, the drop-off of the radiation should be exponential at wavelengths lower than the peak and linear at higher wavelengths.

8. At first, in the 1970s, scientists predicted that the temperature anisotropies should be as large as 1 part in 1,000. Thus the temperature in one location in the sky might be 2.76 K, whereas that in a neighboring region might be 2.763 K. Cosmologists based these predictions on theories of galaxy formation in a Universe containing ordinary atomic matter alone. Because the Universe expanded by a factor of 1,000 between last scattering and the present day, the growth of atomic density perturbations during that period could be at most the same factor—1,000. For the density fluctuations to reach unity by the present day (a requirement for the formation of structure), they had to be at least as large as 1 part in 1,000 at last scattering. Because the photons and atoms moved as a single fluid until last scattering, the CMB temperature fluctuations were then also expected to have the same value—1 part in 1,000. Early CMB experiments searched for anisotropies of this magnitude. However, they were not found. As a result, scientists started to worry that the gravitational instability picture of structure formation might be wrong. Even worse, some wondered whether the lack of anisotropies was telling us about inadequacies in the Hot Big Bang model itself.

Theorists are clever. The lack of temperature anisotropies at the level of 1 part in 1,000 in the data turned out to be reasonable after all. Scientists had forgotten to take into account the role of dark matter in making their predictions.

Computer simulations began to show that, without dark matter, galaxies would never have had time to form. Dark matter density fluctuations formed first and had extra time to grow. As soon as dark matter became the dominant component of mass in the Universe, 7,000 years after the Big Bang, the density perturbations began to amplify. The growth factor by the present epoch was 100,000—significantly larger than in a Universe composed only of atomic matter. Only after last scattering, 380,000 years after the Big Bang, could atoms move independently from photons and fall into the dark matter objects. As a consequence of the extra time for the formation of structure, the predicted value for the temperature anisotropies in the CMB dropped to 1 part in 100,000.

9. John is now the principal investigator for the upcoming James Webb Space Telescope, the sequel to the Hubble Space Telescope. I am hoping that this telescope will discover the dark matter–powered stars named dark stars that I proposed in 2007 with my collaborators Paolo Gondolo and Douglas Spolyar.

10. Hancock, S., G. Rocha, A. N. Lasenby, and C. M. Gutierrez. 1998. "Constraints on Cosmological Parameters from Recent Measurements of CMB Anisotropy." *Letters of the Monthly Notices of the Royal Society* 294: 1.

11. Miller, A. D., R. Caldwell, M. J. Devlin, W. B. Dorwart, T. Herbig, M. R. Nolta, L. A. Page, J. Puchalla, E. Torbet, and H. T. Tran. 1999. "A Measurement of the Angular Power Spectrum of the CMB from $l = 100$ to $400$." *Astrophysical Journal Letters* 524: L1.

12. de Bernardis, P., et al. [Boomerang Collaboration]. 2000. "A Flat Universe from High Resolution Maps of the Cosmic Microwave Background Radiation." *Nature* 404: 955.

13. L2 is the second of five solutions (Lagrange points) to the three-body problem found in the eighteenth century by the mathematician Joseph-Louis Lagrange.

14. http://shawprize.org/en/shaw.php?tmp=3&twoid=67.

15. Figure 3.12 and the Planck panel of Figure 3.14 are based on observations obtained with the Planck satellite (http://www.esa.int/Planck), a European Space Agency science mission with instruments and contributions directly funded by the agency's member states, NASA, and Canada.

16. http://www.crafoordprize.se/press/arkivpressreleases/thecrafoordprize2005.5.32d4db7210df50fec2d800018994.html.

17. Since that time Andrea Ghez has won the 2012 Prize in Astronomy together with Reinhard Genzel. They tracked stars moving around the center of the Milky Way and demonstrated that it hosts a supermassive black hole.

18. For years, I thought I had committed a major faux pas. But recently I discovered a photo of Queen Silvia taken in Oslo on June 15, 2005, where she attended the 100-year celebration of the end of the Swedish-Norwegian union. Luckily this is not the sore subject I thought it was.

**FOUR** Big Bang Nucleosynthesis Proves That Atomic Matter Constitutes Only 5% of the Universe

1. For a review, see Schramm, D. N., and M. S. Turner. 1998. "Big Bang Nucleosynthesis Enters the Precision Era." *Reviews of Modern Physics* 70: 303.

2. Historically the ideas of Big Bang nucleosynthesis began with the work of George Gamow and his collaborators Ralph Alpher and Robert Herman. They viewed the early Universe as a nuclear furnace that could "cook the periodic table." The first paper on Big Bang nucleosynthesis was a seminal contribution to Hot Big Bang cosmology: Alpher, R. A., H. Bethe, and G. Gamow. 1948. "The Origin of Chemical Elements." *Physical Review* 73: 803. Gamow decided to add the name of Hans Bethe to make the author list match the first three letters of the Greek alphabet (alpha, beta, gamma). A study of the early history can be found in H. Kragh. *Cosmology and Controversy.* Princeton, N.J.: Princeton University Press, 1996, pp. 295–305, 338–355.

3. We will encounter today's giant accelerator, the Large Hadron Collider at CERN in Geneva, in Chapters 6 and 7.

4. As I indicated in Chapter 1, I first looked for acting classes in downtown Chicago. After I realized acting classes were out, I discovered the cosmology class was an option, but never would have taken it if I hadn't first heard the lecture described in this chapter.

5. Joyce, J. *Finnegan's Wake.* New York: Penguin Books, 1939.

6. Quarks were first proposed independently by Murray Gell-Mann and Georg Zweig.

7. Pauli's letter was sent to a group of physicists meeting in Tübingen in December 1930 and is reprinted here: http://microboone-docdb.fnal.gov/cgi-bin/RetrieveFile?docid=953;filename=pauli%20letter1930.pdf.

8. Deuterium is sometimes called "heavy hydrogen."

9. More precisely, the expansion rate of the universe became faster than the rates for these reactions.

10. Sagan, C. *Cosmos.* PBS television miniseries, 1980.

11. There are slight discrepancies between predictions and measurements of lithium-6 and lithium-7—still an active area of research.

12. Steigman, G., D. N. Schramm, and J. E. Gunn. 1977. "Cosmological Limit to the Number of Massive Leptons." *Physical Review Letters* B 66: 202.

FIVE  What Is Dark Matter?

1. At that time (to quote Keith), "Supersymmetry was starting to be the fashion for extensions of the standard model." We'll explore the role of supersymmetry in dark matter studies later in the chapter.

2. An important ingredient in all baryogenesis models is a variant of CP violation different from the one that led to the proposed axion dark matter discussed earlier. These decaying superheavy GUT particles satisfy this criterion.

3. Typically, this doubles the number of types of dark matter particles. However, there may be a new twist: in some cases dark matter particles and their antipartners can be identical. Such particles, known as Majorana particles (after the Italian physicist of the same name), can annihilate among themselves. Later in the chapter we return to examples of Majorana particles, as these are among the best candidates for the dark matter of the Universe. In general, however, antimatter and dark matter are different concepts entirely.

4. George Fuller of the University of California, San Diego, and Alex Kusenko of the University of California, Los Angeles, have suggested interesting ways to search for evidence of sterile neutrinos.

5. Bahcall, J. N., C. Flynn, A. Gould, and S. Kirhakos. 1994. "M Dwarfs, Microlensing, and the Mass Budget of the Galaxy." *Astrophysical Journal Letters* 435: L51; Graff, D. S., and K. Freese. 1996. "Analysis of a Hubble Space Telescope Search for Red Dwarfs: Limits on Baryonic Matter in the Galactic Halo." *Astrophysical Journal Letters* 456: L49.

6. The theorists who argued for this extrapolation were Fred Adams and Gregory Laughlin. Fred (my ex-husband) works on the subject of star formation. One day, in an elevator at Columbia University, I introduced him to a well-known high-energy experimentalist. Fred was looking hip in a black leather jacket. I introduced him as "my husband, the star former." The experimentalist misunderstood. He thought Fred was a Hollywood producer.

7. We used data from 114 nearby stars from the U.S. Naval Observatory star catalog.

8. Around the same time Isabelle Baraffe and Gilles Chabrier came to the same conclusions.

9. http://www.nobelprize.org/nobel_prizes/physics/laureates/1983/.

10. All these experiments were created by friends of mine. I can't decide whether to laugh or be enraged at the names they gave their experiments. It is somewhat apt that MACHO was the name of the American group, whereas EROS was that of the French group.

11. Arlin was discouraged from pursuing this line of research when he was a postdoctoral fellow. Hence the first paper written on the subject was based on independent theoretical work by Bogdan Paczynski.

12. This is the same gravitational lensing that was explained in Chapter 2. There its use as a tool for locating dark matter is emphasized.

13. Here I am adapting the words of Madonna from the movie, *Desperately Seeking Susan*. When I gave a talk at the International Center for Theoretical Physics in Trieste on the subject of dark matter, I used this quote. Afterward Alvara de Rújula, at the time the head of the theory group at CERN in Geneva, came up to me and said, "I am a WIMP." When the conference was over, fellow attendee Paolo Gondolo and I drove back to Munich. We had to decide whether to prioritize yet one more meal at our favorite restaurant, the Barcolana, or to leave quickly and see the Dolomites (part of the Alps) before it got dark. We chose the fish at the Barcolana.

14. Glanz, J. February 29, 2000. "In the Dark Matter Wars, WIMPs Beat MACHOs." *New York Times*. I was thrilled to have my photo on the front page of the *Science Times*. Unfortunately it was a bad photo, and nobody even recognized me.

15. http://www.crafoordprize.se/press/arkivpressreleases/thecrafoordprizeinmathematics2012andthecrafoordprizeinastronomy2012.5.4018c179913483dc064280001363.html.

16. A great website illustrating images from the black holes seen by HST is http://www.hubblesite.org/go/blackholes/.

17. Bernard Carr and Stephen Hawking pioneered the work on primordial black holes.

18. My former postdoctoral fellow Daniel Chung together with Rocky Kolb and Antonio Riotto proposed WIMPZILLAs.

19. Rabindra Mohapatra and Vigdor Teplitz investigated mirror matter as a MACHO dark matter candidate.

20. Dimopoulos, S., D. Eichler, R. Esmailzadeh, and G. D. Starkman. 1990. "Getting a Charge out of Dark Matter." *Physical Review* D 41: 2388; de Rújula, A., S. L. Glashow, and U. Sarid. 1990. "Charged Dark Matter." *Nuclear Physics B* 333: 173.

21. The possibility of the lightest Kaluza-Klein particle as a dark matter candidate was introduced by Geraldine Servant and Timothy (Tim) Tait: Servant, G., and T. M. P. Tait. 2003. "Is the Lightest Kaluza-Klein Particle a Viable Dark Matter Candidate?" *Nuclear Physics B* 650: 391.

SIX  The Discovery of the Higgs Boson

1. Though named after Peter Higgs at the University of Edinburgh, the particle was actually independently proposed by six physicists. The others were Robert Brout and François Englert of Université Libre de Bruxelles; Carl Hagen of the University of Rochester; Gerald Guralnik of Brown University, Providence; and Tom Kibble of Imperial College, London.

2. http://www.youtube.com/watch?v=j50ZssEojtM.

3. An electromagnetic calorimeter made of 80,000 crystals of lead tungstate measures the energies of photons and electrons via their electromagnetic interactions. A hadron calorimeter measures the energies of protons, neutrons, pions, and other particles via their predominantly strong interactions.

4. For a nice discussion of the origin of the proton mass, see the article by Frank Wilczek, "The Origin of Mass," in the Massachusetts Institute of Technology's *Physics Annual 2003:* http://www.frankwilczek.com/Wilczek_Easy_Pieces/342_Origin_of_Mass.pdf.

5. Lederman, L., with D. Teresi. 1993. *The God Particle: If the Universe Is the Answer, What's the Question?* Boston: Houghton Mifflin.
And God said, "Let there be mass."

6. Chatrchyan, S. et al. [CMS Collaboration]. 2012. "Observation of a New Boson at a Mass of 125 GeV with the CMS Experiment at the LHC." *Physics Letters* B 716: 30; Aad, G. *et al.* [ATLAS Collaboration]. 2012. "Observation of a New Particle in the Search for the Standard Model Higgs Boson with the ATLAS Detector at the LHC." *Physics Letters* B 716: 1.

7. Cho, A. 2012. "The Discovery of the Higgs Boson." *Science 338: 1524.* http://www.sciencemag.org/content/338/6114/1524.full.

8. http://www.nobelprize.org/nobel_prizes/physics/laureates/2013/.

SEVEN  The Experimental Hunt for Dark Matter Particles

1. Freese, K. September 28, 2000. "The Grid: A Computer Web for Astrophysics and More." *Space.com.*

2. http://home.web.cern.ch/cern-people/opinion/2013/06/how-internet-came-cern.

3. Quote taken from article: Overbye, D. March 29, 2008. "Asking a Judge to Save the World." *New York Times.* http://www.nytimes.com/2008/03/29/science/29collider.html?_r=1&refer=science1).

4. http://www.thedailyshow.com/watch/thu-april-30-2009/large-hadron-collider.

5. After obtaining his PhD in physics from Princeton University, Mark Goodman moved to the Kavli Institute for Theoretical Physics in Santa Barbara, where we were both postdoctoral fellows. Then he switched to science policy, with a focus on the subject of disarmament. In 1990 he won a MacArthur Foundation Fellowship in International Peace and Security at Harvard University, and currently he works at the State Department in Washington, D.C. Physics training can lead to many interesting careers, science policy among them. While Mark and I were both in Santa Barbara, I became friends with Mark's wife Susan Land, who is an excellent fiction writer. I worked on several short stories with her as my teacher, and this book owes a lot to the writing I did back then under her tutelage. I also like to take some credit for their son Jeremy Goodman: a month after I became pregnant with my son Douglas Quincy Adams, Susan apparently felt inspired and became pregnant as well. Our sons have known each other their whole lives.

6. In the 1940s, the physicist Richard Feynman proposed a diagrammatic representation of the interactions of fundamental particles, which has come to be known as the "Feynman diagram." These diagrams are very helpful for computing the rates for such interactions.

7. The original DARKSUSY computer code was written by Gondolo, P., J. Edsjö, P. Ullio, L. Bergström, M. Schelke, and E. A. Baltz. 2004. "DarkSUSY: Computing Supersymmetric Dark Matter Properties Numerically." *Journal of Cosmology and Astroparticle Physics* 7: 8.

8. Although astronomers believe there is some deviation from sphericity, with one axis longer than the other two so that the Galaxy looks more like a football, the standard Halo approximation produces results good to within 10%.

9. The density profile (the fall-off of dark matter with distance from the center) in a dark matter halo of any size is known as a Navarro, Frenk, and White (or NFW) profile, named after scientists Julio Navarro, Carlos Frenk, and Simon White, who wrote the computer simulations.

10. http://www.swpc.noaa.gov.

11. The original goal of the experiment was to search for neutrinoless double beta decay. Here two neutrons would convert to two protons plus two electrons, without emitting any neutrinos. This process would only be possible if a neutrino has nonzero mass. Neutrino mass has now been established by other experiments (as discussed in Chapter 5), yet neutrinoless double beta decay has never been observed.

12. Ahlen, S. P., F. T. Avignone, R. L. Brodzinski, A. K. Drukier, G. Gelmini, and D. N. Spergel. 1987. "Limits on Cold Dark Matter Candidates from an Ultralow Background Germanium Spectrometer." *Physics Letters* B 195: 603.

13. Unfortunately, Peter Smith was forced to retire at 65, at the time the retirement age in England. He now lives in Los Angeles, where I see him every other year at the UCLA Dark Matter Meeting.

14. Early WIMP dark matter theorists included Howie Baer, Lars Bergström, Alessandro Bottino, Manuel Drees, Joakim Edsjo, John Ellis, Josh Frieman, Graciela Gelmini, Paolo Gondolo, Kim Griest, Jim Gunn, Mark Kamionkowski, Rocky Kolb, Benjamin Lee, Dmitri Nanopoulos, Keith Olive, Joel Primack, Leszek Roszkowski, Dave Schramm, Joe Silk, Gary Steigman, Michael Turner, Piero Ullio, Steven Weinberg, and Frank Wilczek; but I can't possibly name them all, so I'll refer to the review articles that list many of the references. Today many more theorists (including Nima Arkani-Hamed, Gianfranco Bertone, Jonathan Feng, Doug Finkbeiner, Dan Hooper, Gordy Kane, Joachim Kopp, Mariangela Lisanti, Aaron Pierce, Stefano Profumo, Pearl Sandick, Tracy Slatyer, Tim Tait, Jay Wacker, Neil Weiner, and Kathryn Zurek) are examining the data and writing down models to try to explain the host of unexplained data that might be due to dark matter. I apologize to my friends not on the list but refer the reader to the following reviews. Jungman, G., M. Kamionkowski, and K. Griest. 1996. "Supersymmetric Dark Matter." *Physics Reports* 267: 195; Bertone, G., D. Hooper, and J. Silk. 2005. "Particle Dark Matter: Evidence, Candidates and Constraints." *Physics Reports* 405: 279; Lewin, J. D., and P. F. Smith, 1996. "Review of Mathematics, Numerical Factors, and Corrections for Dark Matter Experiments Based on Elastic Nuclear Recoil." *Astroparticle Physics* 6: 87; Primack, J. R., D. Seckel, and B. Sadoulet. 1988. "Detection of Cosmic Dark Matter." *Annual Reviews of Nuclear and Particle Science* 38: 751.

15. Silk, J., K. Olive, and M. Srednicki. 1985. "The Photino, the Sun and High-Energy Neutrinos." *Physical Review Letters* 55: 257.

16. Freese, K. 1986. "Can Scalar Neutrinos or Massive Dirac Neutrinos Be the Missing Mass?" *Physics Letters* B 167: 295; simultaneously with Krauss, L. M., M. Srednicki, and F. Wilczek. 1986. "Solar System Constraints and Signatures for Dark Matter Candidates." *Physical Review* D 33: 2079.

17. Gondolo, P., and J. Silk. 1999. "Dark Matter Annihilation at the Galactic Center." *Physical Review Letters* 83: 1719.

EIGHT  Claims of Detection

1. Drukier, A., K. Freese, and D. Spergel. 1986. "Detecting Cold Dark Matter Candidates." *Physical Review* D 33: 3495.

2. The dark matter halo is thought to rotate, but only very slowly.

3. Freese, K., J. Frieman, and A. Gould. 1988. "Signal Modulation in Cold Dark Matter Detection." *Physical Review* D 37: 3388.

4. http://www.nobelprize.org/nobel_prizes/physics/laureates/1921/index.html.

5. This quote is from Frank Avignone. As described in Chapter 7, he headed the first experimental search for WIMPs.

6. http://people.roma2.infn.it/~dama/web/home.html.

7. Graciela Gelmini, Paolo Gondolo, and Chris Savage have written many careful papers on the comparison across experiments. A reference to one paper I wrote

with these authors is: Savage, C., K. Freese, P. Gondolo, and D. Spolyar. 2009. "Compatibility of DAMA/LIBRA Dark Matter Detection with Other Searches in Light of New Galactic Rotation Velocity Measurements." *Journal of Cosmology and Astroparticle Physics* 909: 36.

8. Frank Calaprice's research program investigates fundamental symmetries in nuclear physics. Currently he is also working with postdoctoral fellow Emily Shields to build a new dark matter experiment SABRE to compare with the DAMA results.

9. The University of Michigan is collaborating on PandaX, but any major prize would go to the group leaders from China.

10. Leo Stodolsky, one of the originators of the field of dark matter searches, together with Franz Proebst and Wolfgang Seidel, has been a driving force in this experiment.

11. http://www.sciencenews.org/view/generic/id/349712/description/Dark_matter _detector_reports_hints_of_WIMPs.

12. Sadly, the meeting in 2014 was moved to UCLA campus, because we physicists were simply priced out of the Marriott in Marina del Rey.

13. Glanz, J. February 26, 2000. "Experiments at Stanford Shake Dark-Matter Claim." *New York Times.*

14. Moskowitz, C. October 30, 2013. "Dark Matter Still Hiding: Latest Experimental Sweep Comes Up Empty." *Scientific American.* http://www.scientificamerican .com/article.cfm?id=lux-dark-matter-null-result.

15. Work of Leszek Roszkowski and others.

16. Buckley, M., K. Freese, D. Hooper, D. Spolyar, and H. Murayama. 2010. "High-Energy Neutrino Signatures of Dark Matter Decaying into Leptons." *Physical Review* D 81: 16006.

17. Gondolo, P., and J. Silk. 1999. "Dark Matter Annihilation at the Galactic Center." *Physical Review Letters* 83: 1719.

18. Other members of FERMI at Stockholm University include Jan Conrad, the former convener of FERMI's dark matter efforts.

19. Sarah Kennedy was a cast member of the television show *Saturday Night Live.*

20. My collaborators on the MCubed project are David Gerdes from the Physics Department and Rachel Goldman from the Department of Materials Science and Engineering. We are testing the rate and energy of the gold (and other nuclei) required to break a DNA strand in an ion implementation machine. We have many other experimental issues to study. DNA tends to curl up, but we need it to hang straight down. Magnetic or electric fields can be used to straighten the strands. We want the DNA to be attached in a well-ordered polka dot array, and techniques must be developed to do this. The scooping out of the broken DNA strands will require new ideas, such as using a magnetizable rod.

**NINE** Dark Energy and the Fate of the Universe

1. Astronomers use the word "metal" for all chemical elements heavier than hydrogen and helium, which are the two predominant constituents of the atomic matter in the Universe.

2. This quote is taken from Kirshner, R. 2010. "Foundations of Supernova Cosmology," in *Dark Energy—Observational and Theoretical Approaches,* edited by P. Ruiz-Lapuente. Cambridge: Cambridge University Press.

3. In addition, surveys of relatively nearby (low-redshift) supernovae provided standardized supernova light curves to allow accurate distant measurements.

4. Two groups of astronomers simultaneously discovered that the Universe is accelerating: Riess, A. G., et al. [High-$z$ Supernova Search Team]. 1998. *Astronomical Journal* 116: 1009; Perlmutter, S., et al. [Supernova Cosmology Project]. 1999. *Astrophysical Journal* 517: 565.

5. Specifically, the ratio of the intrinsic versus observed wavelength is defined to be $1 + z$, where $z$ is the redshift. The value of $1 + z$ indicates how compact the Universe was at the time the light was emitted, in comparison to today. At a redshift of $z = 1$, the Universe was twice as dense as it is now; at a redshift of $z = 2$ it was three times as dense, and so on. Higher values of $z$ refer to earlier times in the history of the Universe.

6. The astrophysicists with the foresight to take seriously the data requiring a cosmological constant included Lawrence Krauss, Jim Peebles, Ed Turner, and Mike Turner (no relation to Ed).

7. Weinberg shared the Nobel Prize for the electroweak theory with Sheldon Glashow and Abdus Salam.

8. Freese, K., and M. Lewis. 2002. "Cardassian Expansion: A Model in Which the Universe Is Flat, Matter Dominated, and Accelerating." *Physics Letters* B 540: 1.

9. This work is described in Randall, L. *Warped Passages*. New York: Ecco, 2005.

10. Deffayet, C., G. R. Dvali, and G. Gabadadze. 2002. "Accelerated Universe from Gravity Leaking to Extra Dimensions." *Physical Review* D 65: 44023.

11. When I was younger, I was opposed to physicists using their skills to develop bombs. However, as Matt Lewis pointed out, the military is going to use them anyhow, and isn't it better that they hit their targets rather than taking out hundreds of nearby civilians?

12. Josh Frieman (my collaborator on the rho-vacuum project described previously in the chapter) is the director of DES. The University of Michigan is playing an important role in DES, including Dave Gerdes, Gus Evrard, Dragan Huterer, Wolfgang Lorenzon, Timothy McKay, Jeff McMahon, Chris Miller, Michael Schubnell, and Gregory (Greg) Tarle.

13. Several approaches have been proposed to distinguish a constant from a time-dependent vacuum. With Yun Wang, I investigated the idea of looking directly for time dependence of the vacuum energy density in existing and upcoming data. Other approaches search for a time-changing $w$. Ruth Daly and George Djorgovski champion the idea of searching for a time-changing Hubble constant.

14. A wonderful discussion of the long-term future of the Universe can be found in Adams, F., and G. Laughlin. *The Five Ages of the Universe*. New York: Touchstone Books, 1999.

15. Dyson, F. J. 1979. "Time without End: Physics and Biology in an Open Universe." *Reviews of Modern Physics* 51: 447.

16. Krauss, L. M., and G. D. Starkman. 2000. "Life, the Universe, and Nothing: Life and Death in an Ever Expanding Universe." *Astrophysical Journal* 531: 22; see also Barrow, J. D., and F. Tipler. *The Anthropic Cosmological Principle.* Oxford: Oxford University Press, 1986.

17. From Ball, P. August 3, 2002. "Never Say Die." *New Scientist,* p. 28.

## Afterword: Dark Stars

1. Spolyar, D., K. Freese, and P. Gondolo. 2008. "Dark Matter and the First Stars: A New Phase of Stellar Evolution." *Physical Review Letters* 100: 051101.

# SUGGESTIONS FOR FURTHER READING

Of the many wonderful popular books that have been written about cosmology and related topics, here are a few of my favorites.

Carroll, Sean. *From Eternity to Here*. New York: Dutton, 2010.

Davies, Paul. *How to Build a Time Machine*. New York: Penguin Books, 2001.

Einstein, Albert. *Relativity*. New York: Wings Books, estate of Albert Einstein, 1961.

Freeman, Ken, and Geoff McNamara. *In Search of Dark Matter*. New York: Praxis Publishing, 2006. Note: This book describes the history of astronomical observations that provided compelling evidence for the existence of dark matter in galaxies.

Garfinkle, David, and Richard Garfinkle. *Three Steps to the Universe: From the Sun to Black Holes to the Mystery of Dark Matter*. Chicago: University of Chicago Press, 2008.

Gates, Evalyn. *Einstein's Telescope*. New York: W. W. Norton and Company, 2009.

Greene, Brian. *The Elegant Universe*. New York: Vintage Books, 2000.

———. *The Hidden Reality*. New York: Knopf, 2011.

Guth, Alan. *The Inflationary Universe*. Reading, Mass.: Addison-Wesley, 1997.

Hawking, Stephen. *A Brief History of Time*. New York: Bantam Books, 1998.

Hewitt, Paul G. *Conceptual Physics*. New York: Addison-Wesley, ninth edition, 2001. Note: The chapters on Special and General Relativity are a wonderful introduction to Einstein's work and require no math background at all.

Hooper, Dan. *Dark Cosmos*. New York: Harper-Collins, 2006.

Kirshner, Robert. *The Extravagant Universe*. Princeton, N.J.: Princeton University Press, 2002.

Kolb, Rocky. *Blind Watchers of the Sky*. New York: Basic Books, 1996.

Krauss, Lawrence. *The Physics of Star Trek*. New York: Basic Books, 2007.

Levin, Janna. *How the Universe Got Its Spots*. Princeton, N.J.: Princeton University Press, 2002.

Pais, Abraham. *Subtle Is the Lord—The Science and Life of Albert Einstein*. Oxford: Oxford University Press, 1982.

Primack, Joel, and Nancy Abrams. *The View from the Center of the Universe*. New York: Riverhead Books, 2006.

Rees, Martin. *Our Cosmic Habitat*. Princeton, N.J.: Princeton University Press, 2001.

Riordan, Michael, and David Schramm. *The Shadows of Creation*. New York: W. H. Freeman and Company, 1991. Note: This book is out of print, but used copies may still be available online.

Siegfried, Tom. *The Bit and the Pendulum*. New York: John Wiley and Sons, 2000.

———. *Strange Matters*. New York: Berkley Books, 2002.

Silk, Joseph. *The Infinite Cosmos*. Oxford: Oxford University Press, 2006.

Taylor, Edwin F., and John A. Wheeler. *Spacetime Physics*. New York: W. H. Freeman and Company, second edition, 1992. Note: I learned Special Relativity from this book. The required background is a physics knowledge of force and energy.

Tyson, Neil DeGrasse, and Donald Goldsmith. *Origins: Fourteen Billion Years of Cosmic Evolution*. New York: W. W. Norton and Company, 2004.

Weinberg, Steven. *The First Three Minutes*. New York: Basic Books, 1993. Note: This short book is still the best introduction to the physics of the early Universe.

Wilczek, Frank. *The Lightness of Being*. New York: Basic Books, 2008.

# INDEX

Page numbers for entries occurring in figures are followed by an *f* and those for entries in notes, by an *n*.

Doppler peak, 54, 55, 56, 57f, 60
DRIFT detector, 175
Drukier, Andrzej, 131–33, 131f, 136–40,
    147–49, 175–76
"dunkle Materie," 9
dust, gas, or rocks, dark matter as, 83–84
Duvall, Robert, 2
dwarf galaxies, 144–45, 166, 169, 171–72
Dyson, Freeman, 209–10

Earth: annual modulation due to revo-
    lution around Sun, 138, 147–55, 148f,
    151f, 153f, 173–74; centrality to Uni-
    verse, xii, 211; indirect detection of
    dark matter annihilation products in,
    143–44, 169
Eddington, Sir Arthur, 92
EDELWEISS, 179, 180
EGRET, 170
Einstein, Albert: black holes, 95;
    cosmological constant, 197–98;
    developments in cosmology and, 35;
    gravitational lensing, 20–21, 34; grav-
    ity waves, 97–98; homogeneity and
    isotropy of Universe and, 37; Hub-
    ble's expanding Universe observa-
    tions and, xii, 36–38; modification
    of equations as means of explaining
    dark energy, 195, 204–7; Newton's
    laws and, 28–29; photoelectric effect,
    description of, 149; shape of Universe
    and, 42–43; Special Relativity, 2, 208;
    static Universe, idea of, xi, 37–38, 197;
    warped geometry, 35–36; wormholes,
    99. *See also* General Relativity
Einstein Ring, 22f
electromagnetic calorimeters, 113f, 227n3
electromagnetic force, 70f, 72, 103–4,
    122, 124, 130, 137, 144, 168, 200, 202
electromagnetic radiation, 45
electron degeneracy pressure, 92
electron-positron pairs, 86f, 143, 145–47,
    166, 195, 215
electron volts, 88
electrons: CESR (Cornell Electron Stor-

age Ring), 67–69; density fluctuations
    and, 51; Higgs Boson imparting mass
    to, 116; in ionized universe, 30, 45; last
    scattering, 46, 47f, 50, 51; as leptons,
    69, 81; recombination, 45–46, 46f
electroweak force, 86, 122, 127, 200, 231n7
*Elegant Universe, The* (Greene), 212
elements, formation of, 73, 74–79, 75f,
    76f
Elizabeth II (Queen of England), 128
elliptical galaxies, 13
Englert, François, 116, 122, 227n1
epicycles, 210–11
EROS microlensing experiment, 93,
    226n10
escape velocity from Milky Way Gal-
    axy, 137
Euclid satellite, 206
EURECA, 179
European Organization for Nuclear
    Research. *See* CERN
European Space Agency, 25, 57f, 58f, 60f,
    206, 223n1, 224n15
evaporation of black holes, 98–99
event horizon of black hole, 95–96
evidence for dark matter, 9–34, 211;
    amount of atomic matter in Uni-
    verse, 60, 61, 65–66, 66f, 67, 79–82,
    80f, color figure 3.18; Bullet Cluster,
    27–29, 27f, color figure 2.14; forma-
    tion of galaxies and clusters, 29–32,
    33f, color figure 2.15; galaxy shapes
    and dynamics, 11–13, 12f; gravitational
    lensing, 20–25, 20f, 21f, 22f, 24f, 25f,
    34, color figure 2.11; hot gas in clus-
    ters, xii, 9, 25, 26f, 34, color figure 2.13;
    observational evidence for, 13–29;
    rotation curves, 13–20, 14f, 17f, 18f,
    28–29; Zwicky's study of Coma Clus-
    ter and, 9–11, 10f
Evrard, Gus, 231n12
expansion of Universe: Hubble's obser-
    vations of, xi–xii, 37–39, 37f, 90, 188,
    197. *See also* acceleration of Universe;
    Hubble Expansion

extra dimensions, 106, 204–5
*Extravagant Universe, The* (Kirshner), 195

faint stars, as contribution to Galactic dark matter, 90
Farrar, Glennys, 212f
fate of life in Universe, 209–10
Feinberg, Gary, 136
Fermi Bubble, 171
FERMI satellite, 146, 167, 170–72, 230n18
Fermilab, 2–3, 4, 5, 69, 70, 85, 109
fermions, 115, 146f
Feynman, Richard, and Feynman diagrams, 134, 228n6
Fields, Brian, 93
Figueroa-Feliciano, Enectalí, 160
filaments of large-scale structure, 32, 33f
"fine-tuning," 197
Finkbeiner, Douglas, 171, 172
*Finnegans Wake* (Joyce), 69
flat geometry of Universe, 40–43, 41f, 42f, 44, 54, 55, 57f, 60, 61, 81, 100, 209, 211, 223n3
Flückiger, François, 128
Flug, Sarah Kennedy, 177, 230n19
foosball, 132
Ford, Kent, xii, 17, 18, 19
four fundamental forces of nature, 103, 104
Fred Lawrence Whipple Observatory, Arizona, 170
Freedman, Wendy, 223n1
Freeman, Ken, 19
Freeman, Morgan, 106
Freese, Katherine: appendectomy, recovery from, 1–2; Big Bang nucleosynthesis, first exposure to, 67–69; Cardassian expansion, modification of equations as means of explaining dark energy, 195, 204–5; at Craafoord Prize awards ceremony, 62–64, 62–64f; dark stars, 30–31, 97, 215–18, 222n14; DNA dark matter detectors,

175–79, 177f, 182, 230n20; honorary doctorate from Stockholm University, 172–73; indirect detection, development of ideas for, 143; MACHOs as dark matter and, 90–91, 93–94, 94f; motorcycle gang, rescue from, 192–94; path to studying dark matter, 1–8, 100–101; photos, 5f, 62f, 64f, 133f, 212f, 213f, color figure 9.7; Reines and, 71; Swedish royal family and, 63–65; on *Through the Wormhole with Morgan Freeman* (TV show), 106–7; on time-dependent vacuum energy, 202–3; at Venice Beach and Muscle Beach, California, 161–62; WIMP detector, development of ideas for, 130–34, 136–40, 147–49; at World Science Festival, 212–13, 212f, 213f
Frenk, Carlos, 62f, 137, 228n9
Friedmann, Alexander, 36–37, 40
Frieman, Joshua, 5, 203, 231n12
Fuller, George, 226n4
future evolution of Universe, 207–10; Big Chill 208–9; shrinking horizon and, 207–8; ultimate fate of life, 209–10

Gaia, 58
Gaitskell, Rick, 161f, 162, 165, 179
Galactic Center, 12f, 16, 18, 19, 96, 137, 138, 143, 144, 155, 169, 170–73, 221n4, 222n8
Galactic disk, 11–13
Galactic halo, 13, 18, 18f
galaxies: black holes at centers of, 11, 13, 96–97, 99; components of, 11–13, 12f; density fluctuations leading to, 51; disks in, 11–13; dwarf galaxies, 144–45, 166, 169, 171–72; number of, 211; radio galaxies, 187–88; rotation curves in, 13–20, 14f, 17f, 18f, 28–29; spherical halos of dark matter, 13, 18, 18f; Zwicky galaxies, 79. *See also* Milky Way Galaxy *and other specific galaxies by name*

Martel, Hugo, 200
mass-to-light ratio, 145
Massachusetts Institute of Technology, 22, 160
MAT/TOCO experiment, 54
Mather, John, 52f, 224n9
matter-antimatter asymmetry, 86; and baryogenesis, 86, 114
Max Planck Institute, Munich, 131, 157, 172
Mayer, Maria Goeppert, 136
Mazur, Pawel, 202
McAlpine, Kate, 110
McKay, Timothy, 231n12
McMahon, Jeff, 231n12
MCubed program, 178, 230n20
Mercury (planet), precession of, 28
mercury, Tycho Brahe possibly poisoned by, 16
Messier, Charles, and Messier objects, 39–40
metals, astronomers' use of, 83, 186, 188, 230n1
Mexican Spotted Owl, 170
microlensing experiments, 93–94
MicrOMEGAs, 134
Milgrom, Mordehai, 28–29
Milky Way Galaxy: black hole at center of, 12, 96, 99, 144, 170, 224n17; components of, 11, 12f, 13; detection of dark matter and, 125f, 133, 137–38, 143, 144, 166, 169, 170; disk, 11–13, 12f; evidence of dark matter and, 11–13, 12f, 17, 19, 20, 31, 32; Galactic Center, 12f, 16, 18, 19, 96, 137, 138, 143, 144, 155, 170–73, 221n4, 222n8; halo, 12f, 13; MACHOs in, 90, 93, 94, 99; Sagittarius A*, 12, 96, 99; spiral structure, 11–13, 12f; white dwarfs in, 184
Miller, Amber, 54
Miller, Chris, 231n12
mini-black holes, CERN proton-proton collisions producing, 129
minihalos, 30–31, 215

mirror matter, 101, 103, 227n19
missing transverse momentum and jets, identifying dark matter via, 124–27, 127f
Mobile Anisotropy Telescope, Chile, 54
moguls, on ski slopes, 4, 221n3
Mohapatra, Rabindra, 227n19
MOND (Modified Newtonian Dynamics), 19–20, 28–29
Moore's law, 180
Mottola, Emil, 202, 203
multiverse, 200–202
Muon Spectrometer, 112–13
muons, 69, 81, 85, 113, 116, 119f
Murayana, Reina, 180

NaI (sodium iodide) crystals, 149, 151, 156, 159, 180
NAIAD, 179, 180
Napoleon Bonaparte, 64
NASA (National Air and Space Administration), 23, 27, 48, 50, 57, 58f, 86, 171, 206, 208, 224n15
National Medal of Science, 18
National Oceanic and Atmospheric Organization, 139
Navarro, Julio, 137, 228n9
nebulae, 39f, 40
neon, 164, 165, 182
neutralinos, 105, 145
neutrino mass, 3, 88–89, 228n11
neutrinoless double beta decay, 228n11
neutrinos, 69–72, 70f, 85, 86, 87–90, 113, 131–32, 144–46, 168–69, 180–81, 208
neutron stars, 10, 90, 92–93, 98, 168, 221n2
neutrons: in Big Bang nucleosynthesis, 67; elements, formation of, 73, 74–79, 75f, 76f; isotopes, 74–75; quark structure of, 69, 71f, 72
Newton, Isaac, and Newtonian physics, 14, 19, 28–29, 222n8
NFW (Navarro, Frenk, and White) profile, 137, 228n9
NGC 3198 (galaxy), 18f

primordial black holes, 97, 100
primordial elements, 73, 74–79, 75f, 76f
primordial nucleosynthesis. *See* Big Bang
  nucleosynthesis
Princeton University, 48, 54, 56, 85, 116,
  132, 155, 165, 187, 228n5
Proebst, Franz, 230n10
proton mass, 76, 115–16
Proton Synchrotron, 110
Proton Synchrotron Booster, 110
protons: in Big Bang nucleosynthesis,
  67; CERN accelerator (*See* CERN);
  elements, formation of, 73, 74–79, 75f,
  76f; quark structure of, 69, 71f, 72;
  recombination, 45–46, 46f
protostellar clouds, 216
Ptolemy, 210–11
pulsars, 98, 167–68

Q-balls, 103
QCD (quantum chromodynamics), 101,
  102, 116
quants, 132
quantum field theory, 196f, 198, 205
quantum fluctuations of spacetime, 98,
  99
quark-hadron transition, 73
quarks, 69, 70f, 71f, 72–73, 81, 86, 115,
  116, 145, 225n6
quasars, 22, 79, 83
Queen Silvia of Sweden, 63–65, 63f, 64f,
  224n18

R-parity, 106
radio galaxies, 187–88
"raisin bread model" of Universe, 39
Randall, Lisa, 161f, 205
Ratra, Bharat, 203
recipe for cosmic cocktail, vi
recombination, 45–46, 46f
red giants, 91
redshift, 32, 33f, 37, 40, 188–89, 190–91f,
  231n5
Rees, Sir Martin, 62, 62f
Reines, Frederick, 71

relativity theory. *See* Einstein, Albert,
  and relativity theory
rho vacuum, 203
rich cluster, 9, 24, 25, 27, 44
Riess, Adam, 191, 192, 194f, 231n4
Riotto, Antonio, 227n18
Roberts, Morton, 19
Robertson, Howard, 36–37, 40
Rocha, Graca, 54
Rockefeller University, 132
rocks, gas, or dust, dark matter as,
  83–84
Roman Catholic Church, 211
ROSAT satellite, 25
ROSEBUD, 179
Ross, Diana, 2
Roszkowski, Leszek, 230n15
rotation curves in galaxies, 13–20, 14f,
  17f, 18f, 28–29
Royal Swedish Academy of Sciences,
  62–64f, 62–65, 96
Rubin, Vera, xii, 17, 18, 19f

SABRE, 179, 230n8
saddle-shaped (hyperboloid) geometry of
  Universe, 40–43, 41f, 44, 54
Sadoulet, Bernard, 140, 158, 159f
Sagan, Carl, 78
Sagittarius A*, 12, 96, 99
Sagittarius Galaxy, 144
Saks Fifth Avenue, 173
Salam, Abdus, 231n7
Salkind, Louis, 132
Sano, Takeshi, 176
*Saturday Night Live* (TV show), 230n19
Savage, Christopher, 107, 152, 229–30n7
scalar fields, 203
scalar neutrinos, 144
Schmidt, Brian, 192, 194f
Schramm, David, 4–6, 5f, 69, 81, 84, 88,
  100, 192, 221n3
Schubnell, Michael, 231n12
scintillation, 135f, 149, 158, 164, 179
2dF Galaxy Redshift Survey, 40
Segue dwarf galaxy, 145

Weinberg, Steven, 4, 102, 131, 200, 231n7
Weniger, Christoph, 172
Wetterich, Christof, 203
Wheeler, John, 1–2
white dwarfs, 90, 91–93, 94, 115, 184–86
White, Simon, 137, 228n9
Whitehurst, Robert, 19
Wilczek, Frank, 102, 116
Wilkinson, David, 48, 56–57
Willman, Beth, and Willman I, 145
Wilson, Robert, 48
WIMP miracle, 104, 106
WIMP wind, 138, 148, 174, 181
WIMPs (Weakly Interacting Massive Particles), 103–7; annual modulation of WIMP signal, 138, 147–55, 148f, 151f, 153f, 173–74; as dark matter, 94f, 100, 101, 103–7; defined, 103; extra dimensions, 106; at Harvard, 7–8; head/tail symmetry of, 174, 178; in human body, 106–7; light WIMPs, 166; SUSY (supersymmetry) particles, 104–6, 105f, 121, 126–27, 134, 136, 144, 179. *See also* detection of dark matter particles
WIMPZILLAs, 101, 102–3, 227n18
winos, 105
Witten, Edward, 132

WMAP Haze, 171
WMAP (Wilkinson Microwave Anisotropy Probe) satellite, 54, 56–58, 58f, 59f, 60–61, 132, 171, 191, color figure 3.14
women in physics, 1, 3, 8, 19f, 22, 22f, 44, 63, 132, 136, 149, 156, 157f, 161f, 163–64, 180, 212f, 213
World Science Festival, New York City, 212–13, 212f
World Wide Web, invention of by Sir Timothy Berners-Lee, 128
wormholes, 99

XENON, 152, 153f, 154, 156, 158, 159, 163–65, 179, 213
XMASS, 179

$Z^0$ particles, 70f, 72, 82, 116–18, 119f, 125f, 145
ZEPLIN (British Zoned Proportional Scintillation in Liquid Noble Gases), 141, 150, 165
Zurek, Kathryn, 160
Zweig, Georg, 225n6
Zwicky, Fritz, xii, 9–11, 10f, 13, 25, 79, 221n1
Zwicky galaxies, 79